Microdevices and Microsystems for Cell Manipulation

Special Issue Editors

Aaron T. Ohta

Wenqi Hu

MDPI • Basel • Beijing • Wuhan • Barcelona • Belgrade

MDPI

Special Issue Editors

Aaron T. Ohta   Wenqi Hu
University of Hawaii at Manoa   Max-Planck-Institut für Intelligente Systeme
USA   Germany

Editorial Office
MDPI AG
St. Alban-Anlage 66
Basel, Switzerland

This edition is a reprint of the Special Issue published online in the open access journal *Micromachines* (ISSN 2072-666X) from 2016–2017 (available at: http://www.mdpi.com/journal/micromachines/special_issues/cell_manipulation).

For citation purposes, cite each article independently as indicated on the article page online and as indicated below:

Author 1; Author 2. Article title. *Journal Name* **Year**, *Article number*, page range.

First Edition 2017

ISBN 978-3-03842-618-9 (Pbk)
ISBN 978-3-03842-619-1 (PDF)

Table of Contents

About the Special Issue Editors

Aaron T. Ohta received his B.S. degree from the University of Hawai'i at Manoa, Honolulu, in 2003, his M.S. degree from the University of California at Los Angeles, in 2004, and his Ph.D. degree from the University of California, Berkeley, in 2008, all in electrical engineering. In 2009, he joined the University of Hawaii at Manoa, where he is currently a Professor of Electrical Engineering. He has authored or co-authored over 100 publications in the areas of microelectromechanical systems (MEMS), microfluidics, and reconfigurable electronics. Dr. Ohta was the recipient of the 2012 University of Hawaii (UH) Regent's Medal for Excellence in Research, the ten-campus UH System's most prestigious research award, as well as the 2015 UH Regent's Medal for Excellence in Teaching.

Wenqi Hu received his B.S. degree from the University of Electronic Sci. & Tech. of China, Chengdu, in 2009 and his Ph.D. degree from the University of Hawaii at Manoa in 2014, all in electrical engineering. In 2014, he joined the Max-Planck-Institut für Intelligente Systeme, where he is currently a postdoc researcher. He has authored or co-authored 28 publications in the areas of microrobotics, microfluidics, and reconfigurable electronics. Dr. Hu is currently a fellow of Alexander von Humboldt Foundation and working in the field of biomedical microrobotics.

Preface to "Microdevices and Microsystems for Cell Manipulation"

Microfabricated devices and systems capable of micromanipulation are well-suited for the manipulation of cells. These technologies are capable of a variety of functions, including cell trapping, cell sorting, cell surgery, and cell culturing, often at single-cell or sub-cellular resolution. The functionalities enabled by these microdevices and microsystems are relevant to many areas of biomedical research, including tissue engineering, cellular therapeutics, drug discovery, and diagnostics. This Special Issue of Micromachines, entitled "Microdevices and Microsystems for Cell Manipulation", contains 11 papers (seven articles, three reviews, and one letter) highlighting recent advances in the field of cellular manipulation.

In the areas of cell trapping, cell sorting, and cell characterization, Yousuff et al. review many types of microfluidic systems that are capable of sorting cells based on various characteristics [1]. This review includes cell sorters that use labels or surface markers, as well as the cell's size, shape, deformability, density, compressibility, electrical properties, and magnetic properties to distinguish between cells within a sample. Work by Wang et al. also aimed at characterizing cells. An atomic force microscope was used to mechanically deform cells, and the authors use this information to develop a dynamic model of the cell's viscoelastic properties [2]. This information can also be used to distinguish between various cell types. The trapping and transportation of single cells using optically controlled microbubbles was demonstrated by Fan et al. [3]. This system uses an optically generated thermocapillary force to trap the cells. Dai et al. also used opto-thermally generated bubbles, but vibrated the bubbles to enhance the trapping force [4]. The trapping and transportation of multiple micro-objects using a single bubble was demonstrated, as well as the transportation of mammalian cells and small multicellular algae.

This issue has five papers in the field of cell surgery. Chan et al. used electrical current as feedback to improve the automated microinjection of cells [5]. The effectiveness of the method on neuronal cells was verified by monitoring the injection process and studying the ion channel activities. Another two works on automated cell surgery are from Xie et al. Their first article discusses a visual-servo microrobotic system with both cell autofocusing and an adaptive visual processing algorithm [6], and achieved a 100% microinjection success rate on zebrafish embryos. Their second letter focuses on the challenge of cell fixation during automated cell injection [7]. They used a PDMS microarray cylinder to control the contact force between cells and the material, achieving a microinjection success rate of over 80%. Cell lysis is also a crucial process to extract useful information from the cell interior. In the work of Fan et al. [3], laser-induced microbubbles and their associated microstreams were used to rupture targeted cell membranes and perform single-cell lysis. Other various cell lysis techniques, from macro- to microscale, are reviewed by Islam et al. [8]. This review paper describes the advantages and disadvantages of each technique and compares cell lysis methods applicable to microfluidic platforms.

This issue also contains two papers in the field of cell culturing. Graphene oxide (GO) is suitable for cell growth due to its feature of high hydrophilicity and protein absorption. Kim et al. used meniscus-dragging deposition to fabricate GO micropatterns that affect the distance, speed, and directionality of cell migration [9]. Cell culturing in three-dimensional (3D) scaffolds is expected to significantly impact the fields of drug-screening and tissue engineering. In the work of Liu et al., electrodeposition was used to synthesize alginate hydrogel microstructures in arbitrary shapes, and the assembly of 3D hydrogel blocks was achieved [10]. This platform offers a way for researchers to synthesize complex 3D hydrogel structures for use in tissue engineering. Finally, in the review paper by Vadivelu et al., the use of cell spheroids to culture 3D tissue is surveyed and discussed, along with directions for future development in this area [11].

We would like to thank all the authors for their submissions to this Special Issue. We also thank all the reviewers for dedicating their time and helping to ensure the quality of the submitted papers.

References

1. Yousuff, C.M.; Ho, E.T.W.; Hussain, K.I.; Hamid, N.H.B. Microfluidic Platform for Cell Isolation and Manipulation Based on Cell Properties. *Micromachines* **2017**, *8*, 15, doi:10.3390/mi8010015.

2. Wang, B.; Wang, W.; Wang, Y.; Liu, B.; Liu, L. Dynamical Modeling and Analysis of Viscoelastic Properties of Single Cells. *Micromachines* **2017**, *8*, 171, doi:10.3390/mi8060171.

3. Fan, Q.; Hu, W.; Ohta, A.T. Localized Single-Cell Lysis and Manipulation Using Optothermally-Induced Bubbles. *Micromachines* **2017**, *8*, 121, doi:10.3390/mi8040121.

4. Dai, L.; Jiao, N.; Wang, X.; Liu, L. A Micromanipulator and Transporter Based on Vibrating Bubbles in an Open Chip Environment. *Micromachines* **2017**, *8*, 130, doi:10.3390/mi8040130.

5. Chan, F.H.L.; Yang, R.; Lai, K.W.C. Development of the Electric Equivalent Model for the Cytoplasmic Microinjection of Small Adherent Cells. *Micromachines* **2017**, *8*, 216, doi:10.3390/mi8070216.

6. Xie, Y.; Zeng, F.; Xi, W.; Zhou, Y.; Liu, H.; Chen, M. A Robot-Assisted Cell Manipulation System with an Adaptive Visual Servoing Method. *Micromachines* **2016**, *7*, 104, doi:10.3390/mi7060104.

7. Xie, Y.; Zhou, Y.; Xi, W.; Zeng, F.; Chen, S. Fabrication of a Cell Fixation Device for Robotic Cell Microinjection. *Micromachines* **2016**, *7*, 131, doi:10.3390/mi7080131.

8. Islam, M.S.; Aryasomayajula, A.; Selvaganapathy, P.R. A Review on Macroscale and Microscale Cell Lysis Methods. *Micromachines* **2017**, *8*, 83, doi:10.3390/mi8030083.

9. Kim, S.E.; Kim, M.S.; Shin, Y.C.; Eom, S.U.; Lee, J.H.; Shin, D.-M.; Hong, S.W.; Kim, B.; Park, J.-C.; Shin, B.S.; et al. Cell Migration According to Shape of Graphene Oxide Micropatterns. *Micromachines* **2016**, *7*, 186, doi:10.3390/mi7100186.

10. Liu, Y.; Wu, C.; Lai, H.S.S.; Liu, Y.T.; Li, W.J.; Shen, Y.J. Three-Dimensional Calcium Alginate Hydrogel Assembly via TiOPc-Based Light-Induced Controllable Electrodeposition. *Micromachines* **2017**, *8*, 192, doi:10.3390/mi8060192.

11. Vadivelu, R.K.; Kamble, H.; Shiddiky, M.J.A.; Nguyen, N.-T. Microfluidic Technology for the Generation of Cell Spheroids and Their Applications. *Micromachines* **2017**, *8*, 94, doi:10.3390/mi8040094.

Aaron T. Ohta and Wenqi Hu

Special Issue Editors

micromachines

MDPI

Review

Microfluidic Platform for Cell Isolation and Manipulation Based on Cell Properties

Caffiyar Mohamed Yousuff *, Eric Tatt Wei Ho *, Ismail Hussain K. and Nor Hisham B. Hamid *

Department of Electrical and Electronics Engineering, Universiti Teknologi PETRONAS, 32610 Tronoh, Malaysia; ismailhussain22@gmail.com
* Correspondence: cmd.yousuf@gmail.com (C.M.Y.); hotattwei@utp.edu.my (E.T.W.H.); hishmid@utp.edu.my (N.H.B.H.); Tel.: +60-1678-50269 (C.M.Y.); +60-1238-17752 (E.T.W.H.); +60-1927-87127 (N.H.B.H.)

Academic Editors: Aaron T. Ohta and Wenqi Hu
Received: 28 July 2016; Accepted: 8 November 2016; Published: 4 January 2017

Abstract: In molecular and cellular biological research, cell isolation and sorting are required for accurate investigation of a specific cell types. By employing unique cell properties to distinguish between cell types, rapid and accurate sorting with high efficiency is possible. Though conventional methods can provide high efficiency sorting using the specific properties of cell, microfluidics systems pave the way to utilize multiple cell properties in a single pass. This improves the selectivity of target cells from multiple cell types with increased purity and recovery rate while maintaining higher throughput comparable to conventional systems. This review covers the breadth of microfluidic platforms for isolation of cellular subtypes based on their intrinsic (e.g., electrical, magnetic, and compressibility) and extrinsic properties (e.g., size, shape, morphology and surface markers). The review concludes by highlighting the advantages and limitations of the reviewed techniques which then suggests future research directions. Addressing these challenges will lead to improved purity, throughput, viability and recovery of cells and be an enabler for novel downstream analysis of cells.

Keywords: microfluidics; cell isolation; cell manipulation; cell properties; lab on chip

---

1. Introduction

The human body has abundant cells of different types circulating in bodily fluids. Researchers are discovering a wealth of information about an individual's state of health by analyzing the quantity, quality and type of cells present in these fluids. Selective isolation of cells by property or type from a heterogeneous mixture is the key enabler for a range of useful analysis, i.e., clinical diagnostics, monitoring, therapeutics or as a precursor to biomolecular analysis to understand complex functional interactions between tissues, organs and systems. For example, isolation of rare cells such as circulating tumor cells (CTCs) enriches cellular concentrations to an accurately measureable quantity thereby enhancing accuracy to stratify cancer patients or monitor treatment efficacy [1,2]. Isolation of circulating fetal cells contains genetic information of the fetus, which is useful to identify fetal autosomal abnormalities [3]. Isolation of sperm cells permits manipulations for in vitro fertilization [4] or forensic investigation [5]. Isolation of stem cells permits controlled differentiation of cells to repair or replace injured or diseased tissue [6] in stem cell-based therapies and research. Isolated red blood cells (RBC) and white blood cells (WBC, leukocytes) are required for haematological tests to diagnose a range of illnesses, including anaemia, infection, leukaemia, myeloma and lymphoma. Cancer biomarkers can be analyzed from clean plasma separated from blood for early cancer detection [7]. As isolation accuracy and throughput continues to improve, we anticipate that cellular level therapies will become feasible, i.e., the removal or rehabilitation of dysfunctional or diseased cells such as Human Immunodeficiency

Virus (HIV)-infected T cells or malaria infected RBCs or the removal of pathogenic bacteria circulating in poisoned blood.

To move beyond proof of concept prototypes, these diverse applications of selective cell isolation require high accuracy and reproducibility [8]. Conventional cell isolation systems such as fluorescence-activated cell sorter (FACS), magnetic activated cell sorting (MACS) and centrifugation systems have demonstrated high robustness, accuracy and throughput and have high utility in industrial and lab settings. However, several limitations hinder deployment in more novel applications, such as the requirement for large sample volumes, high reagent consumptions, cross contamination of samples and expensive equipment cost. These systems achieve high throughput cell separation by labelling cells with surface markers or by separating cells based only on density property. To overcome these limitations, researchers devised various cell sorting technologies in miniaturized microfluidic platforms [9–12]. These new technologies enable selective isolation of cells using various other properties like size, shape, deformability, morphology, electrical properties, magnetic properties, compressibility, and also novel surface markers.

Thus far, there is no clear and superior technique to be universally adopted for selectively separating cells. This motivates our review of the different microfluidic mechanisms so that researchers could identify the most suitable technique for their application. We classify the microfluidic techniques by the cellular property employed to distinguish between cells. We believe this grouping is useful because we anticipate that future systems will utilize a suite of cellular features, in tandem or in succession, to achieve greater specificity and reproducibility of separation. Indeed, we discuss several recent innovations towards the paper that have demonstrated this concept. In each group, we highlight the advantages and disadvantages of each technique with respect to device performance and cell viability, throughput, cell recovery rate (yield) and output purity. Recovery rate or capture efficiency expresses the yield of target cells at the outlet compared to all target cells that entered the device. Cell purity expresses the sorting accuracy of sorting and is measured by the percentage of target cells over all cells present at the collection outlet. Cell viability at the most basic level are defined to those cells that are not dead. Throughput (flow rate) determines the number of processed cell per unit time.

2. Based on Induced Cell Properties by Labelled Antibody

2.1. Fluorescence-Activated Cell Sorting

The commercial FACS has become important for biomedical researchers and clinicians for purifying, sorting, counting and analyzing cells which cannot be easily cultured [13–16] such as stem cells, circulating tumor cells and rare bacterial species. The target cells are labelled with antibody-linked fluorescent dye. FACS systems separate target cells via a two-step process. An optical detection system first identifies the presence of fluorescent markers (which have been tagged to target cells only) in a stream of cells. Whenever a target cell is identified, the cell is electrostatically deflected to a collection reservoir by charging the individual cell encapsulated in a droplet. State-of-art FACS perform single-cell separations at rates of 50,000–100,000 cells/s and are capable of distinguishing between 14 and 17 different fluorescent markers [17,18]. There are many commercially available FACS systems which have been reviewed in detail by Picot et al. [19]. FACS systems require high capital investment, high reagent consumption, are only capable of binary separation, require lysing of red blood cells to enhance capture efficiency and may have cross contamination [20].

Microfluidic lab-on-chip devices circumvent these limitations through miniaturization. Instead of electrostatic charging and deflection, a variety of switching mechanisms employing different physics is used to manipulate and sort individual cells. Fu et al. [21] developed the first micro-fabricated FACS device using microfluidic valves for sorting and achieved a throughput of 20 cells/s. Grad et al. [22] designed an X-type junction device powered by two micro solenoid valves at the sheath inlets to separate labelled and unlabeled Fibroblasts at a throughput of 10 cells/min. Chen et al. [23] used a hydrodynamically gated valve. The target cells are separated by gating one outlet with negative pressure and achieved a throughput of 1.4 cells/s.

Sugino et al. [24] invented a novel switch which reversibly transitions from liquid to gel (solid) at elevated temperatures. A thermo-reversible gelation polymer (TGP) solution is mixed into the sample fluid containing cells. When a targeted cell is detected, the liquid mixture is heated to transition the TGP into a gel and deflect the target cell to a different outlet at a throughput of 2.8 cells/s. Target cells may be separated from other cells, but cannot be isolated from the sol-gel solution.

Perroud et al. [25] separated macrophages and Wang et al. [26] to sorted mammalian cells using an optical force based manipulation. A high-power infrared laser deflects target cells into a collection channel and achieved throughput of 20 cells/s. Chen et al. [27,28] used forces from an expanding vapor bubble to push targeted cells into a collection channel at throughput of 10,000 cells/s. The system creates an explosive laser bubble by tightly focusing a pulsed laser beam through a high NA objective lens to heat the liquid medium, as shown in Figure 1A. The system of Nawaz et al. [29], shown in Figure 1B, utilized acoustic force from 150 μs burst of acoustic wave to deflect targeted cells and achieved a throughput of 1200 events/s.

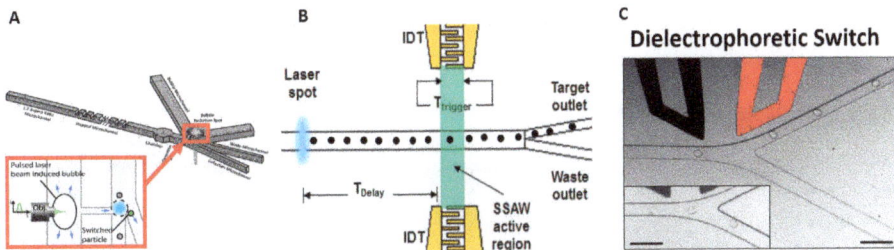

Figure 1. (**A**) Pulsed Laser beam induced bubble to push target cells, Reference [28], Reproduced with permission, Copyright Wiley-VCH Verlag GmbH & Co. KGaA; (**B**) Acoustic radiation force push the cell to target outlet, Reprinted with permission from Reference [29]. Copyright (2015) American Chemical Society; (**C**) Dielectrophoretic force attracts or repels the target cells, Reproduced from Reference [30] with permission of The Royal Society of Chemistry.

Several researchers have encapsulated cells within droplets so that dieletrophoretic force (see Section 6) may be used to push the target cells into the collection outlet [30–33]. The system of Baret et al. [30], shown in Figure 1C, sorted two different strains of *E. coli* and achieved throughput of 2000 cells/s. Target cells were encapsulated in a 12 pL emulsion droplet. Using a similar mechanism but with larger droplet size to improve cell viability, Mazutis et al. [31] demonstrated separation of antibody-secreting cells from non-secreting cells at a lower throughput of 200–400 cells/s.

Mechanically actuated microfluidic FACS systems have low throughput whereas systems actuated by other forces such as acoustic force, bubble expansion and dielectrophoretic force have 10–100× more throughput.

2.2. Magnetic Activated Cell Sorting (MACS)

Magnetic-activated cell sorting (MACS) is another antibody labelled approach similar to FACS. Cells of interest are tagged with marker-specific antibodies conjugated to magnetic labels. The fluid mixture containing tagged and untagged cells is flowed through a strong magnetic field. The magnetically tagged cells are directed into the collection channel by magnetic force. Many commercial extraction kits such as AutoMACS Pro separator (Miltenyibiotec, Bergisch Gladbach, Germany), CELLSEARCH (Janssen Diagnostics, LLC, Raritan, NJ, USA) are available on the market. These kits provide various antibody-labelled magnetic tags for isolation of leukocytes, circulating tumor cells, stem cells, viable cytokine secreting cells, to name a few. These commercial systems can isolate tagged cells with high throughput (10^9–10^{10} cells/h), high purity and high recovery rate but require large samples and labels (magnetic particles), which is costly. Processing is done in batch mode and prolonged duration of operation increases the chance of cross contamination by non-specific binding with the magnetic particles. The review by Hejazian et al. [34] provides more insight into the fundamental physics and important design considerations for MACS systems.

Microfluidics-based magnetic activated cell sorting (µMACS) overcomes these limitations and provides a high purity and recovery rate while requiring fewer magnetic particles with continuous flow. To reduce the volume of magnetic particles needed for cell labeling, microfluidic devices create configurations that elicit stronger magnetic force by increasing the magnetic field gradients crossing the cells, either by increasing the magnetic field strength or increasing proximity between magnetic source and tagged cells. However, there are limitations to the maximum allowable magnetic field gradients imposed by joule heating which reduces cell viability. Various configurations have been implemented using permanent magnets [35–38], electromagnets [39,40], and self-assembled magnets [41]. Osman et al. [42] designed a micromagnet array of Neodymium (NdFeB) films which act as permanent magnet with high magnetic field strength (10^6 T/m).

Many µMACS methods used the H channel structure to separate target cells from a mixture with two inlets and two outlets [43–47], shown in Figure 2A. The mixture of magnetically labeled and non-labeled cells are introduced into one of the inlets and sheath flow is introduced into the other inlet at the same flow rate. Laminar flow in the micro channel keeps the streams distinct and permanent magnets placed beside the streams attract magnetically tagged cells to cross the stream into the collection channel. By optimizing the placement and distribution of magnetic force, Del Giudice et al. [48] achieved up to 96% separation efficiency at flow rate of up to 4 µL/min, using the concept illustrated in Figure 2B. Cells from multiple target groups can be tagged with differently sized magnetic particles and experience different magnetic force and deviations into different outlets [35].

Among the efforts to improve the performance of magnetophoresis system, Forbes et al. [49] developed a comprehensive numerical model as a tool to design magnetophoretic systems. With aid from the tool, they demonstrated separation of magnetically labeled breast adenocarcinoma (MCF-7) cells using lateral magnetophoresis with an angled permanent magnet configuration as shown in Figure 2C. Lee et al. [50] isolated magnetically tagged *E. coli* bacteria from whole blood and improved efficiency to almost 100% at throughput of 60 mL/h by cascading three similar separation stages in series and parallel as shown in Figure 2D.

Kirby et al. [51] combined centrifugation with magnetophoresis to isolate magnetically labelled cells in a device depicted in Figure 2E. Centrifugation forces push cells radially outward along the microchannels as the device spins at high speeds. Magnets close to the channel pull the target cells into a collection chamber while the remaining unlabeled cells are collected at the end of the microchannel. They successfully demonstrated the isolation of MCF-7 cancer cells from whole blood and HIV/AIDS relevant epitopes from the whole blood [52] with high sensitivity of 1 cell per mL at a capture rate of 88% and 92%, respectively.

As an alternative to focusing, Hoshino et al. [53] demonstrated capture of magnetically tagged cancer cells from a continuous flow of blood using a microfluidic device with magnets arrayed at the bottom of the channel. They achieved a sensitivity of 1000 cells down to 5 cells per mL with a capture efficiency of ~86% for two cancer cell lines at a throughput of 10 mL/h. Similarly, Shields et al. [54] created a device with patterned micromagnets over microwells to isolate magnetically labeled CD4+ lymphocytes from blood with 95% accuracy, shown in Figure 2F.

Magnetic-label sorting compares favorably against fluorescent-label sorting because identification and actuation of target cells are performed at the same step. However, the throughput reported in the literature is low (4 µL/min–60 mL/h) to allow sufficient duration for the magnetic force to displace tagged cells into the collection channel. Also, antibodies and the attached magnetic labels are difficult to remove from the surface and can alter the properties of target cells. It remains an open challenge to detach magnetic beads from labelled cells after separation [17].

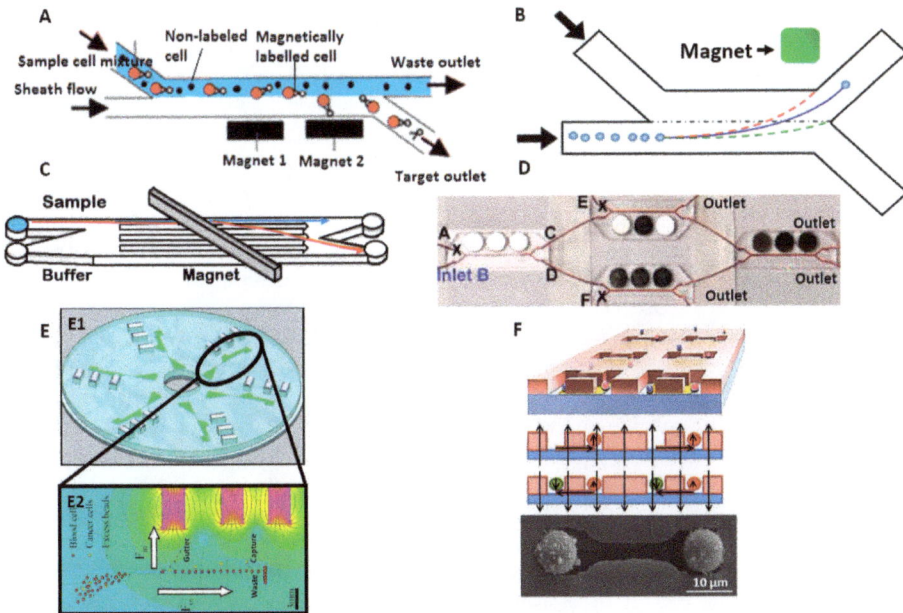

Figure 2. (**A**) A schematic of H filter for magnetic based separation (**B**) Viscoelastic focusing of magnetic particles Reproduced from Reference [48] with permission of The Royal Society of Chemistry; (**C**) Angled permanent magnet configuration Reproduced from Reference [49] with permission of The Royal Society of Chemistry; (**D**) Cascade magnetic separation stages, Adapted with permission from Reference [50]. Copyright (2014) American Chemical Society; (**E**) Schematic of Lab on disc chip with microfluidic channels, visible in green, and magnets as silver (E1) and Inset view of one of the channel in E1, where simulation shows, blood cells, excess beads collected at waste, target cells at capture and bead waste at gutter (E2), Reproduced with permission, Reference [52], Copyright Wiley-VCH Verlag GmbH & Co. KGaA; (**F**) Patterned micromagnets over microwells to isolate magnetically labelled cells, Reproduced from Reference [54], with the permission of AIP Publishing.

3. Separation Based on Cell Surface Markers Properties

Panning is a classical method to isolate cells where a surface is functionalized with antibodies and target cells flowing through the surface are captured and tightly bound by antibody coupling. This technique is frequently used but limited by low capture efficiency and purity levels. To compensate, large number of cells (10^5–10^7) are required which makes panning unsuitable for applications with a small volume of analytes such as small cell populations harvested from organs and tissues or applications with high false positive rates from non-specific bindings [55].

To improve yield purity and capture rate over classical antibody treated surfaces, microfluidic implementations have various innovations in the design of surfaces [56–65]. Conventionally, CTC cells are selectively captured on surfaces functionalized with antigen Epithelial cell adhesion molecule (anti-EpCAM) antibodies [56]. Although expressed in most CTCs, the surface antigen EpCAM is notably absent on some non-epithelial and melanoma cells. Capture efficiency also decreases as flow rate is increased. Gaskill et al. [59] increased capture efficiency up to 98% at a flow rate of 18.7 µL/min by adding a second adhesion molecule, E-selectin, to the microchannel surface. Through rapid bonding and bond-breaking of leukocytes to E-selectin, leukocytes accumulate near the capture surface and impede CTC flow to maximize CTC cell contact with anti-EpCAM on the capture surface.

Kurkuri et al. [60] achieved capture efficiency of cancer cells between 80% and 90% at throughput of 48 µL/min by increasing the area of the capture surface. They created microposts on the polydimethylsiloxane (PDMS) capture surface and functionalized with EpCAM. Mitchell et al. [61] developed an alternative design to increase surface area. They immobilized nanostructured Halloysite nanotubes (HNT)-coated with E-selectin on the capture surface to increase the surface area to capture CTC from leukocytes. They achieved 90% purity at a throughput of 0.04 mL/min.

Captured cells can be released with aptamer functionalized surfaces. Zhu et al. [62] demonstrated reversible capture and cell viability post-release on T-Lymphoblast human cell lines (CCRF-CEM) and Zhang et al. [63] demonstrated selective capture of *E. coli* from a mixture with different bacterial species. Target cells are captured via affinity binding and non-target cells are removed by washing. When the surface temperature is increased (48 °C for 2 min) by micro-heaters arrayed on the functionalized surface, aptamer binding is disrupted and the target cells are released. They were able to capture the target cells of approx. 250 cells/mm^2. Jeon [64] also developed a conductive nano surface to electrically release and retrieve captured CTCs. The capture surface was first electrodeposited with biotin-doped conductive polypyrrole and subsequently functionalized with biotinylated antibodies. Up to six different cell lines of CTCs can be selectively captured by adhesions through the various antibody types. Captured CTCs are successfully released through application of a voltage pattern to the conductive polypyrrole, which liberates the biotin-CTC conjugate for collection. With this method, they achieved capture and release efficiency of 97% and 95%, respectively, at a flow rate of 1.2 mL/h.

Instead of selectively immobilizing target cells, Bussonie et al. [65] demonstrate an alternative principle of separating cell types according to differences in adhesion strength to the surface. They captured HEK293 and A7R5 cells on a surface coated with LiNbO$_3$ and using a surface acoustic wave to detach the adherent cells, they achieved sorting purity of 97% and sorting efficiency of 95%.

Separation by surface marker properties has been popularly applied to isolating CTCs from a heterogeneous mixture of whole blood. However, it remains challenging to improve the selectivity for other cell types because many markers expressed on the cell surface are not unique to a particular cell type.

4. Separation Based on Size, Shape, and Deformability

Cell separation by physical properties such as size, shape, and deformability are promising avenues for real-time diagnostic and clinical applications. This label-free approach does not require expensive chemical reagents or antibody labelling thus reducing sample preparation time while improving throughput and cell viability. The various techniques that utilize these physical properties to isolate cells are discussed below.

4.1. Membrane Filtration (Size)

Membrane filtration in microfluidic devices utilize thin membrane layers with micro-pores of predefined size, geometry and spatial distribution. Target cells, predefined by size, are not permitted to flow through the membrane whereas other cell types and the liquid may pass freely. The membrane filter is made from various polymer material like thermoplastic material [66], parelyene material [67], photo definable material [68], and PDMS membrane [69]. A schematic depicting membrane filtration is shown in Figure 3A. Cell separation by membrane filtration has been demonstrated on numerous cell types; for example, Fatanat et al. [66] isolated rare oligodendrocyte progenitor cells (OPCs) from rat brain tissue, Zheng [67] isolated CTC from whole blood and Adams et al. [68] isolated MCF-7 cell lines from whole blood.

Commercially available CTC separation systems like Parsortix system (ANGLE plc., Guildford, UK) use size-based filtration and achieve a capture rate of 66% and a release rate of 61% at throughput of 0.17 mL/min [70] and fewer depleted leukocytes (200–1000 cells) [71,72]. It outperforms commercial labelled sorting systems like ScreenCell (ScreenCell, Sarcelles, France) [73], IsoFlux (Fluxion Biosciences,

Inc., San Francisco, CA, USA) [74], CellSearch system, and AutoMACS. IsolFlux has a capture rate of 40% whereas CellSearch has a release rate of 19% [70].

Although conceptually simple, careful design of membrane filters is crucial for good performance. To maximize throughput, accuracy and precision of cell separation, membranes must have good uniformity of pore geometry, high porosity [67,75–77] and be thermally and mechanically stable. Proper choice of through-hole diameter is the crucial design parameter to separate the target cells. The membrane fabrication process must give uniform microstructures and researchers have endeavored to reduce fabrication complexity [69,78] to improve quality. Membranes are also optically transparent to facilitate further downstream analysis. Clogging of membrane pores by large particles is a frequent problem which rapidly degrades capture efficiency. If cross-flow injectors are added, clogging can be minimized because non-adherent cells are flushed away from the membrane surface.

Chen et al. [78] demonstrated an approach to control the size of filtrates artificially and separate leukocytes from whole blood. In this system, shown in Figure 3B, leukocytes were conjugated to microbeads and were readily trapped on the PDMS microfiltration membrane (PMM) whereas other non-target cells could pass through the membrane freely. The diameter of through holes in PMM was smaller than the microbeads but greater than unbound leukocytes. The device achieved throughput of 20 mL/min with 97% capture purity. Furthermore, the leukocytes trapped on the PMM could be analyzed with immunophenotyping assays like AlphaLISA [79], i.e., to determine cytokine concentrations. Compared to conventional whole blood stimulation assays, like ELISA (enzyme-linked immunosorbent assay) or ELISpot (Enzyme-Linked ImmunoSpot), this new process yielded a 10-fold reduction in processing time.

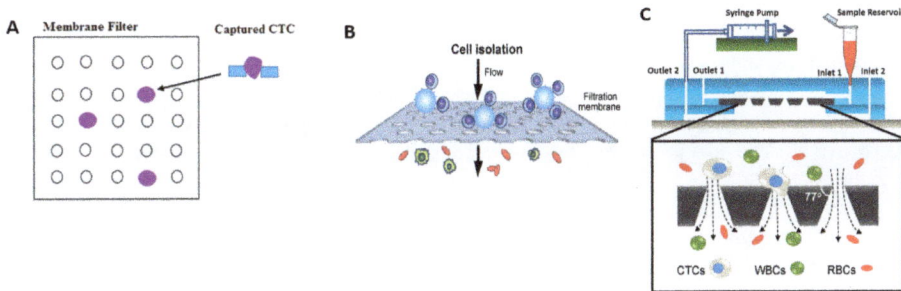

Figure 3. (**A**) A schematic of membrane filtration; (**B**) polydimethylsiloxane (PDMS) membrane filter, Reference [78] Reproduced with permission, Copyright Wiley-VCH Verlag GmbH & Co. KGaA; (**C**) Conical Membrane Filter, Adapted by permission from Macmillan Publishers Ltd.: [Nat. Commun.], Reference [80], copyright (2014).

Similar work also reported recently by Fan et al. [69] for isolation of CTC with a capture rate of >90% and relatively higher processing throughput at 10 mL/h. This method is found to be the best choice for CTC separation and downstream analysis. Other than uniform circular pores, conical pores designed by Tang et al. [80], as shown in Figure 3C, with cross flow components has shown promising results in capturing CTCs, with a high efficiency of 96% and a high purity at flow rate of 0.2 mL/min. In addition to this, captured CTCs can be cultured in the same device for further analyses and to clearly understand the cell behavior, which is generally adopted.

Membrane filtration has disadvantages, i.e., cell proliferation and release from the filter pores are restricted [81]. Also, cells are often washed with fixatives to prevent lysis during filtration [66–68]. Zhou et al. [82] attempted to overcome these limitations by designing a separable bilayer (SB) microfilter to separate viable CTC cells from blood with efficiency of 83% and cell viability of 74% at 0.4 mL/min flow rate. The gap between the two layers and the pore alignment are designed to induce less mechanical stress on CTCs, thus improving cell viability for further analysis downstream.

4.2. Inertial Separation (Size)

Inertial separation utilizes inertial forces within flowing fluid to deflect cells. By careful design of microchannel dimensions and geometry, cells of different sizes will migrate to different positions along the micro-channel induced by various forces, i.e., smaller cells will be dominantly affected by dean forces which are generated by a lateral, secondary-vortex flow along a spiral channel whereas the larger cells will migrate under inertial lift forces. Inertial lift forces arise from two competing forces, namely the shear induced lift, which pushes cells towards the channel wall due to the interaction with the fluid velocity profile, and the wall induced lift, which pushes cells away from the channel wall due to interactions at close proximity to the channel wall. To generate sufficient displacement force, the cell–fluid mixture must flow at a sufficiently high velocity with Reynolds number, Re >> 1 [83]. Di Carlo first showed the utility of inertial forces by demonstrating cell focusing and cell deflection in a variety of different channel geometries such as asymmetrically curved channels [84], straight high aspect ratio channels and contraction-expansion arrays (CEA) [85].

Gregoratto [86] demonstrated separation using spirally shaped microchannels at a flow rate of 2 mL/min. Son [87] isolated non-spherical cells such as sperm cells from RBCs and successfully recovered 81% of non-motile sperm at two outer wall outlets and 99% of RBCs are recovered at two inner wall outlets at a flow rate of 0.52 mL/min with the system shown in Figure 4A. Other groups have demonstrated successful sorting using variations of the spiral configuration for blood plasma, CTCs and malaria-infected cells [88–91]. Jimenez et al. [92] demonstrated separation of viable waterborne pathogens like Cryptosporidium parvum. Hong et al. [93] utilized two 90° curved channels to separate viruses, bacteria and larger aerosols into three outlets, thus showing an improvement over traditional bioaerosol sampling methods.

Following this, Papautsky and Bhagat invented several spiral architectures for continuous multi-particle separation [83,94–98] operating at higher flow rates of up to 3 mL/min. They invented a two-stage separation technique to separate a range of different cell types including blood and tumor cell. The first stage utilizes inertial lift force in low aspect ratio channel to separate cells into two equilibrium streams near the top and bottom walls of the microchannel. Streams from the first stage enter the second stage where rotationally induced lift forces filter cells with enhanced efficiency. The achieved 99% efficiency separating Human Prostate Epithelial (HPET) tumor cell from diluted blood [97]. As shown in Figure 4B, Warkiani invented a multiplexed system with three stacked spiral channels and achieved complete isolation of various blood components like plasma [99], CTCs [99–102], Malaria infected cells [100] and CHO (Chinese Hamster Ovary) cells [99] with this design, achieving a high throughput of 7.5 mL/5 min [88].

Instead of curved geometries, contraction-expansion channels, shown in Figure 4C, can also be utilized for separation as demonstrated by Hur et al. [85]. At high flow rate, microscale vortices develop in the expansion region of the channel to entrap the larger sized cancer cells while smaller sized blood cells continue unhindered in a focused flow. Captured cells are later released by reducing the flow rate through the channel. Cells can be processed at a throughput of upto 7.5×10^6 cells/s at flow rate of 4.5 mL/min. The same group developed a high throughput vortex chip (Vortex HT) [103], by adding two parallel channels and 1.5× more serial reservoirs for cell capture which improved the capture efficiency from 20% to 83%. This design achieves high throughput of 8 mL/min and purity of 85% for separation of MCF-7 breast cancer cells from whole blood.

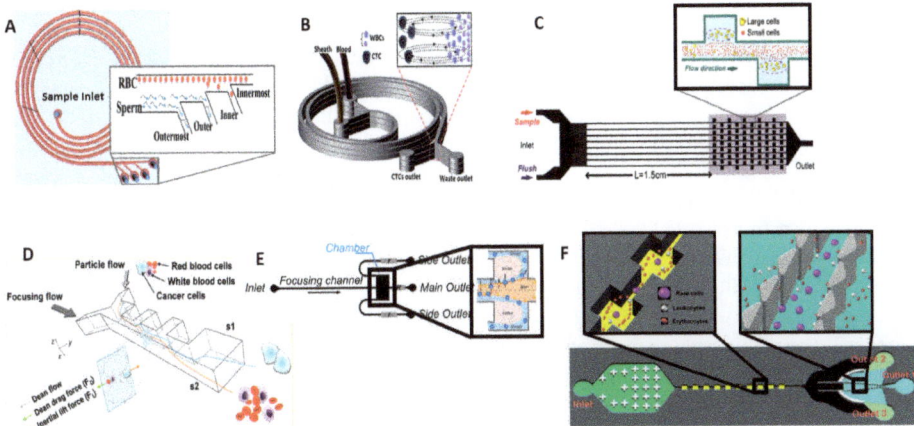

Figure 4. (A) Spiral microchannel for sperm cell and RBC isolation, Reproduced from Reference [87] with permission of The Royal Society of Chemistry; **(B)** Stacked spiral channel for high throughput applications, Reproduced from Reference [102] with permission of The Royal Society of Chemistry; **(C)** Contraction-Expansion with Vortex Aided Separation, to trap larger cells in expanded reservoir region due to the difference in the lift forces on the cells Reproduced from Reference [85], with the permission of AIP Publishing; **(D)** Contraction-Expansion array separation, Reprinted with permission from Reference [104]. Copyright (2013) American Chemical Society; **(E)** Vortex-aided inertial microfluidic with siphoning outlet, Reproduced from Reference [105], with the permission of AIP Publishing; **(F)** Inertial force and Steric hindrance at the outlet for effective separation, Reproduced from Reference [106] with permission of The Royal Society of Chemistry.

Lee et al. [104] achieved a high recovery rate of 99.1% and throughput of 1.1×10^8 cells/min using a two-stage CEA channel, shown in Figure 4D. The design utilized a sequence of high-aspect-ratio channels with alternating large and small widths which causes blood cells to focus close to the channel sidewalls due to inertial lift forces. By selecting the width of the pinching or narrow regions to be on the order of the largest cell size (cancer cells), CTCs are refocused along the channel axis while the rest of the cells remain aligned around the sidewalls. A bifurcating outlet at the final stage funnels the segregated CTCs, red blood cells and peripheral blood leukocytes to different outlet channels. Using only a single CEA channel, Wang et al. [105] created siphoning outlets beside the entrapment chambers, shown in Figure 4E, to continuously draw off 86% of the large cells trapped in the vortices and achieved 99% purity when separating RBCs from large polystyrene particles at a flow rate of 0.5 mL/min. Shen et al. [106] also designed a multistage device by combining inertial microfluidics with a size-based pre filtering with CEA and post filtering with steric hindrances to sort blood cells and tumor cells MCF-7 and HeLa cells, shown in Figure 4F. This combined stage provides >90% recovery rate at throughput of 2.24×10^7 cells/min with >92% viable cells were found.

Zhou et al. [107] studied the design parameters responsible for effective trapping and separation in symmetric, rectangular expansion channel, i.e., dimensions of expansion region, sample concentration, threshold Reynolds number designed for trapping particles at higher flow rate of $Re = 180$, and achieved 98% trapping efficiency.

High throughput and fine cutoff between sizes can be achieved but the challenge remains. Flows populated densely with cells will destabilize the separation efficiency and this places upper limit on maximum achievable throughput. Also, size alone may not be a sufficiently discriminative feature between cell types.

4.3. Deterministic Lateral Displacement (Size, Shape, and Deformability)

The deterministic lateral displacement (DLD) approach separates cells based on differences between the combination of size, shape and deformability. Cells are flowed through an array of micro-structures or micropillars, as shown in Figure 5A, and are differentially displaced when the flow forces cells around the obstructing micro-structures. By carefully designing the positional offset of successive rows of micropillars, cells larger than a designed critical radius will be gradually deviated from the initial flow whereas small cells will flow along the streamline of the initial flow [108].

Researchers investigated the efficacy of different pillar shapes. Circular pillars were popularly studied for separation of various cell types [108–111]. For instance, Holm et al. [108,109] successfully isolated RBCs from diluted blood. Liu et al. [112] subsequently demonstrated that triangular pillars are better suited for cell types with large deformations because cells deform minimally around the pillar apex. They achieved separation efficiency between 80% and 99% for separation of various cancer cell types (MCF-7 and MDAMB231) from blood at a throughput of 2 mL/min. Loutherback et al. [113] achieved a higher throughput of 10 mL/min by a novel arrangement of triangular pillars shown in Figure 5B, which isolates CTCs from blood with capture efficiency greater than 85% and no impediments to cell viability. Zeming [114,115] studied the efficacy of various micropillar shapes i.e., circular, square, I-shaped, T-shaped, L-shaped and anvil, as shown in Figure 5C. The I-shaped pillar was most effective at separating non-spherical from spherical cells. The protrusion and groove structure induces rotational motion on the non-spherical cells thus increasing its effective diameter. The net effect of the I-shaped pillar is to enhance the disparity of sizes between the spherical and non-spherical cells. Another recent innovation is to reduce the downstream gap size between micropillars [116]. This optimization does not restrict throughput but enhances the separation efficiency of RBC up to 100% at a flow rate of 0.2 μL/min from whole blood.

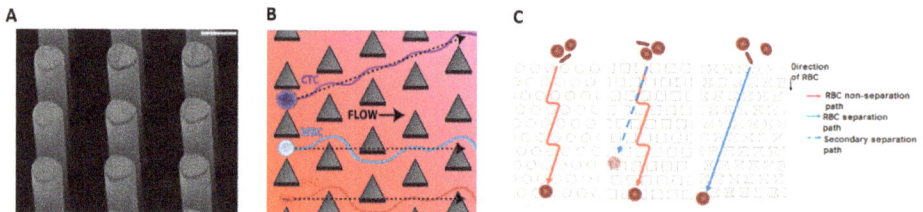

Figure 5. (**A**) Circular Pillar array for white blood cells (WBC) and red blood cells (RBC) separation, Adapted by permission from Macmillan Publishers Ltd.: [Nat. Commun.], Reference [110], copyright (2014); (**B**) Triangular Pillar array for Circulating Tumor Cell (CTC) Separation, Reproduced from Reference [113], with the permission of AIP Publishing; (**C**) Circular, Square, and I-Shaped pillars for RBCs separation, Adapted by permission from Macmillan Publishers Ltd.: [Nat. Commun.], Reference [114], copyright (2013).

DLD devices suffer from several key limitations. Individual pillars are prone to defects in size, shape and height [117] and any structural irregularities within the array tends to blur the streams, degrade separability of cells and increase the possibly of stiction and blockages. Long channel lengths are required to achieve significant lateral displacement [118].

5. Separation Based on Size, Density and Compressibility

5.1. Centrifugation and Pinched Flow Fractionation (Size and Density)

Fluids with a mixture of densities have the natural tendency to sediment into layers of increasing density under the influence of gravitational forces. Centrifugation accelerates this separation process by means of high magnitude centrifugal force generated from high speed rotation. Low et al. [119]

recently reviewed commercial systems that employ centrifugation for CTC isolation. Commercial centrifugation systems are designed with powerful motors to circulate large volumes of liquid.

In contrast, microfluidic implementations of centrifugation are designed to process small volume of fluids between 5 and 2000 μL [120,121]. Several researchers have adapted the Compact Disc (CD) systems to achieve high speed rotational motion with low cost commodity parts. The microfluidic channels are created on the CD and rotated at very precise and consistent speeds on the audio playback system. Li [122] achieved 92% separation efficiency of blood plasma from whole blood at 2000 rpm. Burger [123] also demonstrated up to 80% of plasma extraction within a short, 2 min interval with the system shown in Figure 6A. Lee et al. [124] used centrifugation to push cells in unprocessed whole blood through a membrane filter, depicted in Figure 6B, and successfully isolated CTC with capture efficiency of 61%.

In pinched flow fractionation (PFF), cells are focused in a narrow channel and abruptly enter an expansion area and come under the influence of gravity, drag and buoyancy forces. Cells that are less dense than the surrounding medium will be lifted to the upper regions of the expanded chamber. Cells with higher density will experience greater gravity force and flow towards the bottom of the chamber. This technique was demonstrated for isolation of spherical and non-spherical particle having similar density [125], particles of different size [126], CTC from WBC [127]. Song et al. [128] observed that the height differences between two cells of different densities could be amplified by decreasing the flow rate, depicted in Figure 6C. Morijiri et al. [129] also demonstrated PFF separation but combined with centrifugation to generate a larger and controllable sedimentation force to enhance the separation between particles of different densities at a throughput of 2060 μL/h.

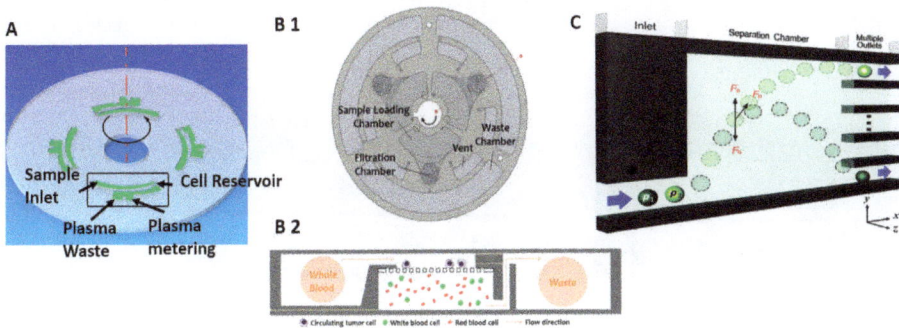

Figure 6. (**A**) The disc containing four identical plasma separation structure, Reproduced with permission from Reference [123], © IOP Publishing. All rights reserved; (**B**) The top view of the CTC-isolation disc showing detailed microfluidic features (B1) and the schematic illustration showing the working principle of the CTC-isolation disc (B2), Reprinted with permission from Reference [124]. Copyright (2013) American Chemical Society; (**C**) Conceptual microfluidic design for density based separation at different heights. Reprinted from Reference [128], with the permission of AIP Publishing.

Many cell types have similar density and size and centrifugation is not able to distinctly separate between them. Though some literature reported the high throughput separation using a centrifugal based approach, still this method is not sufficiently demonstrated for different cell types including rare cells, white blood cells, and whole blood cell mixture.

5.2. Acoustophoresis (Density and Compressibility)

Acoustophoresis based cell isolation is a technique which uses the density and compressibility properties of a cell to distinguish different cell types. For example, CTCs and WBCs have different compressibility [130]. In acoustophoretic microdevices, high-intensity sound waves [131] interact with the microchannel to generate pressure gradients that push cells into specific spatial locations.

The acoustic waves reflect off the microchannel walls to establish a standing wave pattern within the microchannel. Cells flowing through the standing wave are moved towards high pressure or anti-pressure node by radiation forces. The magnitude of the radiation force depends on the volume, density and compressibility of the cell, the surrounding medium and the amplitude and wavelength of acoustic wave. Cells with greater density and compressibility than the surrounding fluid will migrate towards the pressure node. Bands of cells, grouped by density and compressibility, form across the microchannel. Since flow is laminar, cells will hold the position in the band even after crossing the acoustic zone and conveniently collected at separate outlets [132], shown in Figure 7A.

Yang et al. [133] demonstrated separation of apoptotic cells from viable cells of the same type. They achieved 94% recovery of viable cells with 91% purity at throughput of 5 mL/h. Burguillos et al. [134] studied the impact of acoustophoretic separation on cell viability, proliferation and cell response to subtle phenotypic changes. They demonstrated that acoustophoretic processing did not affect cell viability of prostate cancer cells nor the respiratory functions for human thrombocytes and leukocytes.

Ding et al. [135] showed that particles could be displaced towards different outlets by modulating the frequency of a standing surface acoustic wave (SSAW) using the prototype shown in Figure 7B. To improve separation efficiency and increase separation distance between cell types, they tilted the angle [136] of the standing surface acoustic waves (taSSAW) relative to the channel as shown in Figure 7C to generate multiple pressure nodal lines. This configuration increases the likelihood of capturing cells within one of the nodal lines and they achieved a recovery rate of 71% with a purity of 84% when separating cancer cells from blood. By optimizing inclination angle, channel dimension and SSAW input power, they improved the cancer cell recovery rate to 83% at throughput of 20 μL/min [130]. Cell types are separated by displacements between 100 and 250 μm which is significantly larger than the displacements achieved by inertial forces and dielectrophoresis (DEP) [137,138].

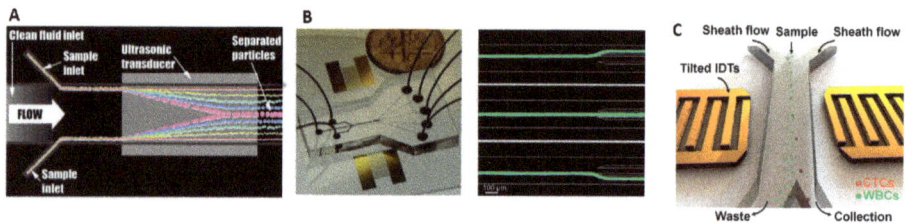

Figure 7. (**A**) Illustration of a suspended particle crossing transducer is pushed to center channel based on their acoustic properties, Adapted with permission from Reference [131]. Copyright (2007) American Chemical Society; (**B**) standing surface acoustic wave (SSAW)-based cell-sorting device with three inlets and five outlets and 15 μm fluorescent particle are directed to three different outlet channels by adjusting the applied frequency (14.5 MHz, SAW off, and 13.9 MHz), Reproduced from Reference [135] with permission of The Royal Society of Chemistry; (**C**) Schematic illustration of a high-throughput the angle of the standing surface acoustic waves (taSSAW) device for cancer cell separation, Reproduced with permission from Reference [130], Copyright (2015) National Academy of Sciences.

Acoustic-based separation has a good recovery rate and high purity but lacks the limitations of operating at low flow rates as discussed in the literature so far. Secondly, even though acoustic based separation was applied to sort CTCs from other cells like WBCs, still compressibility properties of clinical CTCs and other rare cells are largely undocumented and unknown. Therefore, it initiates future studies to establish ideal physical property profiles from different cell populations to optimize the device performance.

6. Separation Based on Electrical Properties

Dielectrophoresis (DEP) is based on the principle that a force is induced on a dielectric particle when it is placed in a non-uniform electric field. Cells are not charged but polarized in the non-uniform field. Polarized cells experience a translational force, called DEP force, which either deflects (negative DEP) or attracts (positive DEP) towards the region of maximum field strength. The polarization acquired by cells depend on the cellular conductivity and permittivity, inherent polarizability of the fluid and also the magnitude and frequency of the applied electric field. To enhance the electric field gradient and the force induced on the cells, liquid solutions with free electrolytes are frequently added to the cell mixture. DEP was applied to sort CTC [139–141], blood cells [142–144], immune cells [145,146] and pathogens [147,148].

A common DEP strategy is to induce deflections of target cells to separate them from a flow. The basic design comprises of a H channel with sample and sheath flow inlets, two outlets and sidewall electrodes that generate the electric field gradient across the microchannel [136,140,144,146]. Thick metal electrodes are needed to generate uniform electric field gradients over the entire height of the microchannel [144,149] but are challenging to fabricate by metal deposition. To overcome these limitations, Lewpiriyawong et al. [137] developed three dimensional Ag-PDMS composite electrodes in the sidewall of the microchannel, shown in Figure 8A. Piacentini et al. [142] demonstrated a design with liquid electrodes, shown in Figure 8B, and successfully isolated RBCs and platelets with 98% purity and recovery rate. To enhance separation distance without undue increases in the electric field amplitude. Researchers also demonstrated optically induced DEP (ODEP) on a photoconductive sheet of amorphous silicon. The ODEP or virtual electrode was created on the sheet by projecting a pattern of light from a commercial projector as shown in Figure 8C. Different electrode geometries could be created in ad hoc fashion by simply changing the pattern of projected light. Researchers have achieved separation of prostate cancer cells (PC3) from leukocytes and viable sperm cells from non-motile cells [150–152] with separation efficiency of >85% at a particle velocity of 150 μm/s.

Song [144] used an array of oblique interdigitated electrodes to separate hMSC cells from osteoblasts and achieved separation efficiency of 92% at throughput of 5.4 μL/min with purity of 97%. The osteoblasts where polarized and deflected by DEP force (positive or negative) following a zig-zag trajectory, shown in Figure 8D. Ling et al. [152] invented a novel method of deflecting target cells from a flow using an array of triangular-shaped electrodes at the base of the microchannel. Due to its geometry, the apex of each electrode has a strong electric field, which attracts target cells by positive DEP. Target cells are incrementally deviated from the main flow by each peak it traverses while the untargeted cells remain undisturbed in the flow. They separated live fibroblast cells (NIH-3T3) from dead cells in Swiss mice, and osteosarcoma cells (MG-63) from erythrocytes with 82.8% separation efficiency at a throughput of 1302 cells/min.

Instead of deflecting cells in a continuous flow, the system designed by Jen et al. [153,154] required the cell fluid mixture to be dropped onto patterned planar electrodes. The electrodes form a concentric circular pattern. By activating and deactivating the electric potential between adjacent electrode pairs, the target cells are gradually moved to the center of the planar electrode by positive DEP force. They separated HeLa cells from MCF-5 cells with separation efficiency above 80%.

An alternative DEP strategy is to trap target cells by DEP forces and later release them for collection. Many researchers opted for array arrangements and novel electrode geometries surrounding the microchannel. Microwell arrays attract and trap target cells inside the wells by DEP forces. The target cells may be subsequently released into a separate collection reservoir by removing the voltage applied to the microwells. The design is similar to microliter well plates but scaled down in size and fabricated by drilling holes in a three-dimensional laminate made from stacked layers of graphene [155] or copper [156] sheet (interleaved with insulating layers). The resulting microwell walls act as electrodes. Another common design using extruded electrodes generate strong DEP forces, which improves throughput and affords better trapping efficiency up to a flow rate of 12 μL/min. Targeted cells such as HL-60 cell [157], polystyrene particles, drosophila cells [158], and viable yeast cells [155,159] are trapped by the electrodes in the microchannel with positive DEP while non-target cells are flushed to the outlet via negative DEP. Arrays of extruded electrode posts have been fabricated from gold [157], carbon [158] and highly doped silicon [160].

New electrode configurations are being invented and some have not yet been tested on live cells. Yafouz et al. [161] demonstrated a novel electrode configuration combining positive and negative DEP to trap polystyrene particles from 1 to 15 μm at separate locations around the electrode. Using a ring-shaped microarray of dot electrodes, the larger size particles were repelled and concentrated in the center of the ring shape by negative DEP while the smaller sized particles were attracted to the edge of the dot by positive DEP. Lapizco et al. [162,163] trapped particles between an array of insulating posts within a microchannel. When a direct current (DC) potential is applied between the inlet and outlet of the microchannel, the insulating posts divert electric field lines through the spaces between the posts, thereby increasing the local electric field strength there. As a result, a strong positive DEP force attracts and traps target cells between the posts.

To generate sufficient forces for deflection or trapping, high electric fields are needed for DEP but it inflicts membrane stress and induces joule heating on cells, which may eventually lead to cell death. Devices with insulating hurdles, as shown in Figure 8E [138], were designed to minimize the deleterious effects of high electric field exposure to cells. The hurdle is a constricted region joining both sides of the channel. The constriction compresses the electric field lines thus creating a localized region of high field strength. Cells experience high electric fields and deflection by DEP forces only within this region, thus minimizing exposure to its dangerous influence. Variations on the hurdle shape have been proposed, i.e., multiple rectangular insulating hurdles [164], triangular hurdles within an H-shaped microchannel [165,166] and S-shaped hurdles [167]. Hurdle designs have been used to separate biological cells [168] and DNA molecules [169].

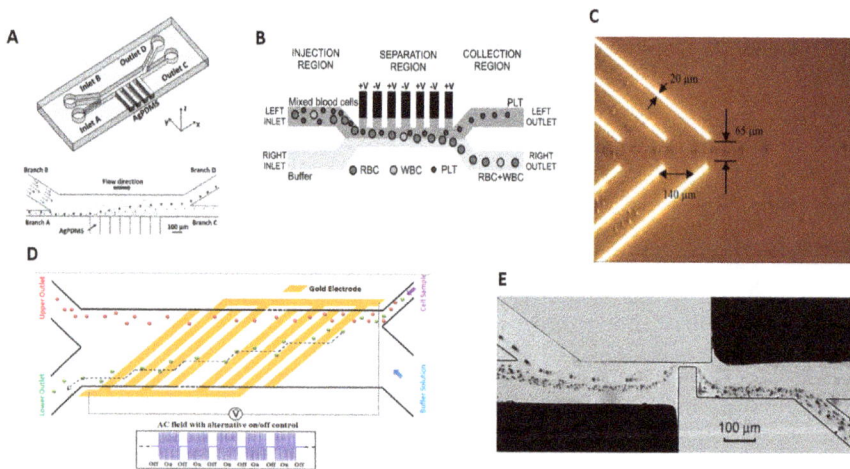

Figure 8. (**A**) PDMS microdevice with 3D sidewall composite electrodes with separation mechanism in the fabricated microdevice, Reference [137], Reproduced with permission, Copyright Wiley-VCH Verlag GmbH & Co. KGaA; (**B**) Liquid Electrodes placed at the left side of the channel for dielectrophoresis (DEP)-based platelets, RBCs and WBCs separation, Reproduced from Reference [142], with the permission of AIP Publishing); (**C**) Cell Focusing using Virtual electrodes (ODEP) Reprint with permission, from Reference [149], Copyright Elsevier (2008); (**D**) An oblique interdigitated electrodes for the continuous flow DEP based microfluidic cell separation, Adapted from Reference [144] with permission of The Royal Society of Chemistry; (**E**) Copper electrode with insulating PDMS hurdles for 5 μm, 10 μm and yeast cell separation Reproduced from Reference [138] with permission, Copyright Elsevier.

However, DEP remains challenging to deploy in a reliable manner because dielectrophoretic forces are highly sensitive to buffer conditions (e.g., salt concentration) and achieve low throughput (typically <100 cells/s per microchannel) [133]. DEP based systems have thus far not demonstrated separation of multiple target cells. Several review articles have recently enumerated the advantages of DEP techniques [170,171].

7. Separation Based on Intrinsic Magnetic Properties of Cells

Unlike Section 2.2 whereby magnetic properties were induced in targeted cells by antibody labelling or ingestion of magnetic nanoparticles, RBC and WBC have intrinsic magnetic properties which can be utilized for separation. Deoxygenated RBCs are paramagnetic which means these cells are attracted towards the source of a magnetic field. In contrast, oxygenated RBCs and WBCs are diamagnetic and therefore deflected from magnetic field source [172,173].

Furlani et al. [172] demonstrated isolation of red blood cells and white blood cells from plasma by embedding a microarray of Nickel-based soft magnetic material around the microchannel. When they magnetized the microarray by bringing permanent magnets into proximity, deoxygenated RBCs were attracted and WBCs deflected into two peripheral outlets and blood plasma exited the middle outlet. Using a similar principle, Nam et al. [173] isolated malaria infected RBCs because the malarial parasite produces the by-product hemozoin which makes the sick RBCs develop paramagnetic properties. The achieved a recovery rate of 73% and 98.3% for early and late stage infections, respectively, and a separation efficiency of 99% at 1.6 μL/min flow rate. Robert et al. [174] demonstrated isolation of macrophages and monocytes based on endocytotic properties which have internalized with magnetic nanoparticles. Monocytes are less endocytotic than macrophage and a permanent magnet placed on top of the chamber deflects and sort the cells based on their iron loading. They were successfully sorted with a purity of more than 88% and an efficiency of more than 60% at flow rate of 50 μL/h.

8. Multi-Target Separation Utilizing Multiple Cell Properties

As discussed so far, each cell sorting technique has its own advantages and limitations. To separate cells with more nuanced differences, multiple cellular properties should be simultaneously employed by combining techniques in order to improve sorting efficiency.

Mizuno et al. [175] demonstrated a hybrid technique to separate two different cell-types, leukocytes (JM cells) and cancer cells (HeLa cells) and bin all cells from each type by size. In their two stage process, they first used a sheath flow to push cells in a focused flow near the microchannel wall. The stream of cells pass through a series of outlets branching out of the main channel with progressively larger openings, as shown in Figure 9A. Cells exit the microchannel through the first outlet that matches their size. At the end of every outlet, a magnet displaces the magnetically tagged JM cells from the HeLa cells into different outlets to yield groups of leukocytes and cancer cells separated into bins according to size. They achieved a sorting efficiency of 90% and throughput of 10 μL/min.

To separate three different strains of *E. coli* bacteria, Kim et al. [176] combined dielectrophoresis and magnephoresis displacement technique, referred to as integrated dielectrophoretic-magnetic activated cell sorter (iDMACS) as shown in Figure 9B. Bacteria from strain A were tagged with polystyrene beads and deflected by DEP forces to one outlet. Strain B was tagged with streptavidin-coated magnetic particles and captured by a magnet whereas Strain C was unlabeled and allowed to flow through the 2nd outlet. They achieved ~98.9% efficiency for both tagged strains and 100% efficiency for the unlabeled strain at throughput of 2.5×10^7 cells/hour in a single pass separation process.

Karabacak et al. [110] invented a three stage process to separate CTC from whole blood, as shown in Figure 9C. In the first stage, WBCs and CTCs were isolated from other blood cells using deterministic lateral displacement. DLD is a very efficient technique to separate the highly deformable RBCs and small-sized platelets. At the next stage, the collected CTC and WBC cells are lined up by inertial focusing and, subsequently, WBCs (which have been magnetically tagged) are separated from CTCs by magnetophoresis. The WBCs were tagged instead of CTCs because antibody tagging may induce cytotoxicity on the CTCs and impede downstream analysis of the isolated cells. Furthermore, some types of CTCs (of epithelial origin) do not express the binding antibody and thus cannot be labeled. Inertial focusing of WBCs and CTCs prior to magnetophoretic separation improves separation efficiency because all cells flowing in a single focused stream will experience very similar magnetic forces. With this combination of techniques, the high throughput action of inertial and DLD separation could be combined with the precision of magnetophoresis separation. Furthermore, viability of the separated cells, namely CTC, is preserved for accurate downstream analysis. They achieved a 97% yield of rare cells with a processing throughput of 8 mL/h of whole blood.

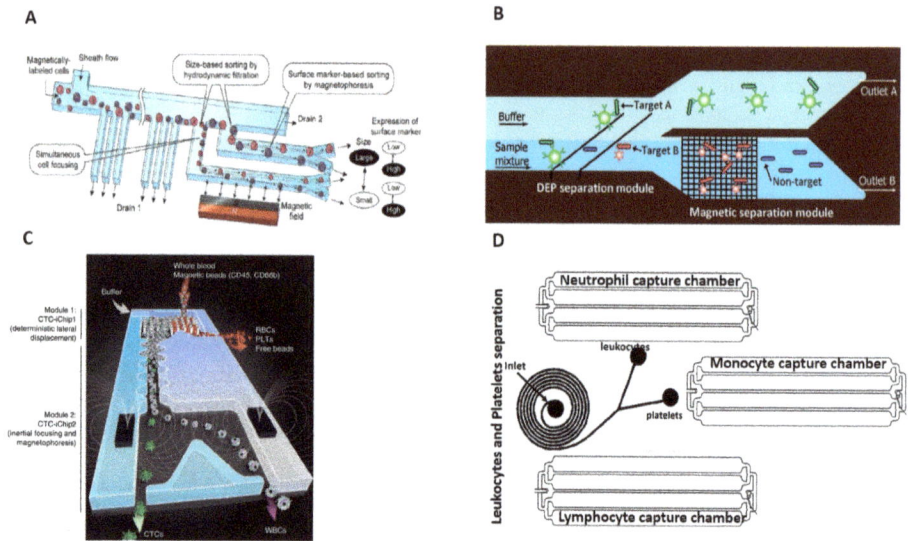

Figure 9. (**A**) Principle of a two-dimensional (2D) cell sorting system integrating hydrodynamic filtration (HDF) and magnetophoresis, Reprinted with permission from Reference [175]. Copyright (2013) American Chemical Society; (**B**) The integrated dielectrophoretic-magnetic activated cell sorter (iDMACS) device architecture with DEP and MACS stage, Reproduced from Reference [176] with permission of The Royal Society of Chemistry; (**C**) Hybrid deterministic lateral displacement (DLD), Inertial and Magnetophoresis device, circulating tumor cells integrated Chip (CTC-iChip) for cancer cell separation, Adapted by permission from Macmillan Publishers Ltd.: [Nat. Protoc.], Reference [110], copyright (2014); (**D**) A schematic of Hybrid Inertial and adhesion based system for leukocytes subtypes.

To separate immature RBCs from the blood of pregnant women, Huang et al. [177] also demonstrated a DLD separation step followed by intrinsic magnetic separation. Immature or nucleated RBCs (NRBCs) are indicators of fetal health and are less deformable than mature RBCs. DLD separates the NRBCs and WBCs from mature RBCs and, at the subsequent stage, a magnetic column deflects the WBCs and attracts the NRBCs into separate outlets. They achieved efficiency of 99% at a throughput of 27 mL/h.

Huang et al. [178] demonstrated a hybrid DEP-immunocapture system to extract prostate cancer cells (LNCaP line) from peripheral blood. They functionalized the device surface to capture antigens expressed by the cancer cells. Through careful tuning of an applied AC electric field, negative DEP forces were generated to repel the non-specific leukocytes while attracting the cancer cells for adhesion through positive DEP forces. This combination of techniques yielded improvements in the capture efficiency up to 85% and purity of <5% leukocytes with a modest throughput of 0.2 mL/h.

Gupta et al. [179] demonstrated separation of multiple cell targets with a single round of separation with increased purity, yield, and throughput. They integrated inertial separation with adhesion sorting in two stages, as shown in Figure 9D, to separate various types of leukocytes from platelets. In the first stage, they used a spiral device to separate leukocytes from platelets. In the second stage, the leukocytes pass through a series of capture channels, each having different leukocyte binding moieties to capture individual leukocyte types like monocytes, neutrophils, and lymphocytes. They achieved purity of 91% and yield of 80% at throughput of 1.5 mL/min.

Hybrid techniques enable more selective and efficient separation of multiple cell targets from a mixture of multiple cell types.

9. Challenges and Future Research Directions

Beyond proof of concept demonstrations, the microfluidic methods reviewed are promising methods to implement new clinical applications. Individually, each method has limitations to be overcome and we now briefly discuss the key challenges for future research.

The antibody labelled approach demonstrates good results in purity, recovery rate, and separation. However, the antibodies remain attached to the labelled cells even after isolation, and are difficult to remove from the surface of the cells. Properties and functions of target cells may be altered by the antibodies and this is motivation to develop methods to detach the antibody, especially for applications that require the target cells to be reintroduced into the human body after analysis, such as differentiated stem cells and other healthy cells. In some designs, it is not clear if the limitations can be solved, i.e., where joule heating from magnetic field switching may denature proteins and irrevocably damage or destroy the cells.

Similar drawbacks also limit isolation based on cell surface marker properties. Target cells captured by adhesion to a functionalized surface are difficult to detach. Further optimizations are necessary to increase the low throughput and also to reduce non-specific binding of cells to fully realize the potential for selecting cell targets accurately from a mixture of many cell types in a single pass.

Where cells have distinct intrinsic electrical properties, precise and selective isolation with high recovery rate can be attained with DEP force by tuning the electric field frequency. However, many cell types have similar dielectric properties and this technique cannot be used to efficiently separate them [180]. Methods that separate cells by density and compressibility also permit selection of target cells through tuning of the acoustic wave frequency. Separability becomes an issue when cell types have a broad range or an overlap of size and densities. Both dielectrophoresis and acoustophoresis systems operate with limited flow rate. Beyond a threshold speed, sorting purity degrades because the induced forces are too briefly applied to the target cells.

In contrast, size-based separation with inertial forces requires higher flow rate. Membrane filtration is another common approach to separate cells by size but the membrane is prone to pore clogging by larger cells and this reduces the efficiency of cell sorting and coalescence of pores leads to less capture efficiency as targeted cells may flow through the pores and not be captured. DLD micro-devices do not require dilution of blood sample but these demonstrations either operate at low flow rate or are susceptible to clogging between the micropillars.

All methods reviewed except inertial devices operate at low throughput. Also, all techniques have demonstrated separation between only two different cell types with dilution. In contrast, the commercial parsortix system® demonstrated successful isolation of CTCs from whole blood with almost no dilution [73].

Utilizing several properties like size, shape and deformability simultaneously leads to improved selectivity of target cells from the heterogeneous cell mixture, even when differences between cell types are subtle. Hybrid techniques combining two or more principles for separating cells may be mutually complementary and overcome limitations of individual techniques. In oncology research, microfluidics plays a vital role to isolate rare cancer cells from the normal cells as a preparatory step for further sequential and downstream analysis. High recovery rate with minimal background contamination (high purity) is critically important for downstream analysis to enable clear and reliable investigation of specific cell types. In addition, high throughput is necessary to isolate a sufficient quantity of these rare cells for processing, typically at least 1–10 mL of raw blood while maintaining the cell viability.

10. Conclusions

Microfluidic devices have numerous advantages over the traditional bench top methods and are emerging as a new platform for enabling applications in clinical diagnostics and therapeutics such as in oncology. Though the microfluidic device is sterile and disposable, the limitation lies in the capability to process a large sample volume. Though there has been progressive improvement to the methods reviewed, there is often competing tradeoffs between recovery rate, purity, throughput and viability for approaches targeting a single cell property. Hybrid techniques utilizing several cellular properties are a promising approach to isolate multiple cell types by exploiting the benefits of multiple cellular properties in a single pass process.

Acknowledgments: We thank Universiti Teknologi PETRONAS for funding Caffiyar Mohamed Yousuff and Ismail Hussain K under the Graduate Assistantship scheme. Eric Tatt Wei Ho thanks the Ministry of Education Malaysia for support under the Higher Institution Center of Excellence funding program.

Author Contributions: C.M.Y. wrote the paper. All authors designed the study and edited the paper.

Conflicts of Interest: The authors declare no conflict of interest.

References

1. Gossett, D.R.; Weaver, W.M.; Mach, A.J.; Hur, S.C.; Tse, H.T.K.; Lee, W.; Amini, H.; Di Carlo, D. Label-free cell separation and sorting in microfluidic systems. *Anal. Bioanal. Chem.* **2010**, *397*, 3249–3267. [CrossRef] [PubMed]
2. Massimo, C. Circulating tumor cells, disease progression, and survival in metastatic breast cancer. *N. Engl. J. Med.* **2006**, *351*, 781–791.
3. Fiddler, M. Fetal cell based prenatal diagnosis: Perspectives on the present and future. *J. Clin. Med.* **2014**, *3*, 972–985. [CrossRef] [PubMed]
4. Janosek-Albright, K.J.C.; Schlegel, P.N.; Dabaja, A.A. Testis sperm extraction. *Asian J. Urol.* **2015**, *2*, 79–84. [CrossRef]
5. Garvin, A.M.; Fischer, A.; Schnee-griese, J.; Jelinski, A.; Bottinelli, M.; Soldati, G.; Tubio, M.; Castella, V.; Monney, N.; Malik, N.; et al. Isolating DNA from sexual assault cases: A comparison of standard methods with a nuclease-based approach. *Investig. Genet.* **2012**, *3*, 25. [CrossRef] [PubMed]
6. Wei, X.; Yang, X.; Han, Z.; Qu, F.; Shao, L.; Shi, Y. Mesenchymal stem cells: A new trend for cell therapy. *Acta Pharmacol. Sin.* **2013**, *34*, 747–754. [CrossRef] [PubMed]
7. Garbett, N.C.; Merchant, M.L.; Helm, C.W.; Jenson, A.B.; Klein, J.B.; Chaires, J.B. Detection of cervical cancer biomarker patterns in blood plasma and urine by differential scanning calorimetry and mass spectrometry. *PLoS ONE* **2014**, *9*, e84710. [CrossRef] [PubMed]
8. Mohamed, H. Use of Microfluidic Technology for Cell Separation. In *Blood Cell—An Overview of Studies in Hematology*; INTECH: West Palm Beach, FL, USA, 2012; Chapter 11.
9. Hardt, S.S. *Microfluidic Technologies for Miniaturized Analysis Systems*; Springer: Berlin, Germany, 2006.
10. Li, D. *Encyclopedia of Microfluidics and Nanofluidics*; Springer: Berlin, Germany, 2008.
11. Hsu, T. *MEMS and Microsystems: Design, Manufacture, and Nanoscale Engineering*; John Wiley & Sons, Inc.: Hoboken, NJ, USA, 2008.
12. Tabeling, P. *Introduction to Microfluidic*; Oxford University Press: Oxford, UK, 2006.
13. Cho, S.H.; Chen, C.H.; Tsai, F.S.; Godin, J.M.; Lo, Y.-H. Human mammalian cell sorting using a highly integrated micro-fabricated fluorescence-activated cell sorter (microFACS). *Lab Chip* **2010**, *10*, 1567–1573. [CrossRef] [PubMed]
14. Gross, A.; Schoendube, J.; Zimmermann, S.; Steeb, M. Technologies for single-cell isolation. *Int. J. Mol. Sci.* **2015**, *16*, 16897–16919. [CrossRef] [PubMed]

15. Corselli, M.; Crisan, M.; Murray, I.R.; West, C.C.; Scholes, J.; Codrea, F.; Khan, N.; Bruno, P. Identification of perivascular mesenchymal stromal/stem cells by flow cytometry. *Cytom. A* **2013**, *83*, 714–720. [CrossRef] [PubMed]

16. Pasut, A.; Oleynik, P.; Rudnicki, M.A. Isolation of muscle stem cells by fluorescence activated cell sorting cytometry. *Methods Mol. Biol.* **2012**, *798*, 53–64. [PubMed]

17. Hosic, S.; Murthy, S.K.; Koppes, A.N. Microfluidic sample preparation for single cell analysis. *Anal. Chem.* **2016**, *88*, 354–380. [CrossRef] [PubMed]

18. Piyasena, M.E.; Graves, S.W. The intersection of flow cytometry with microfluidics and microfabrication. *Lab Chip* **2014**, *14*, 1044–1059. [CrossRef] [PubMed]

19. Picot, J.; Guerin, C.L.; Le, C.; Chantal, V.K. Flow cytometry: Retrospective, fundamentals and recent instrumentation. *Cytotechnology* **2012**, *64*, 109–130. [CrossRef] [PubMed]

20. Autebert, J.; Coudert, B.; Bidard, F.-C.; Pierga, J.-Y.; Descroix, S.; Malaquin, L.; Viovy, J.-L. Microfluidic: An innovative tool for efficient cell sorting. *Methods* **2012**, *57*, 297–307. [CrossRef] [PubMed]

21. Fu, A.Y.; Spence, C.; Scherer, A.; Arnold, F.H.; Quake, S.R. A microfabricated fluorescence-activated cell sorter. *Nat. Biotechnol.* **1999**, *17*, 1109–1111. [PubMed]

22. Grad, M.; Young, E.F.; Brenner, D.J.; Attinger, D. A simple add-on microfluidic appliance for accurately sorting small populations of cells with high fidelity. *J. Micromech. Microeng.* **2013**, *23*. [CrossRef] [PubMed]

23. Chen, P.; Feng, X.; Hu, R.; Sun, J.; Du, W.; Liu, B.-F. Hydrodynamic gating valve for microfluidic fluorescence-activated cell sorting. *Anal. Chim. Acta* **2010**, *663*, 1–6. [CrossRef] [PubMed]

24. Sugino, H.; Ozaki, K.; Shirasaki, Y.; Arakawa, T.; Shoji, S.; Funatsu, T. On-chip microfluidic sorting with fluorescence spectrum detection and multiway separation. *Lab Chip* **2009**, *9*, 1254–1260. [CrossRef] [PubMed]

25. Perroud, T.D.; Kaiser, J.N.; Sy, J.C.; Lane, T.W.; Branda, C.S.; Singh, A.K.; Patel, K.D. Microfluidic-based cell sorting of Francisella tularensis infected macrophages using optical forces. *Anal. Chem.* **2008**, *80*, 6365–6372. [CrossRef] [PubMed]

26. Wang, M.W.; Tu, E.; Raymond, D.E.; Yang, J.M.; Zhang, H.; Hagen, N.; Dees, B.; Mercer, E.M.; Forster, A.H.; Kariv, I.; et al. Microfluidic sorting of mammalian cells by optical force switching. *Nat. Biotechnol.* **2005**, *23*, 83–87. [CrossRef] [PubMed]

27. Chen, Y.; Wu, T.-H.; Kung, Y.-C.; Teitell, M.A.; Chiou, P.-Y. 3D Pulsed laser-triggered high-speed microfluidic fluorescence-activated cell sorter. *Analyst* **2013**, *138*, 7308–7315. [CrossRef] [PubMed]

28. Chen, Y.; Chung, A.J.; Wu, T.; Teitell, M.A.; Carlo, D. Di pulsed laser activated cell sorting with three dimensional sheathless inertial focusing. *Small* **2014**, *10*, 1746–1751. [CrossRef] [PubMed]

29. Nawaz, A.A.; Chen, Y.; Nama, N.; Nissly, R.H.; Ren, L.; Ozcelik, A.; Wang, L.; Mccoy, J.P.; Levine, S.J.; Huang, T.J. Acoustofluidic fluorescence activated cell sorter. *Anal. Chem.* **2015**, *87*, 12051–12058. [CrossRef] [PubMed]

30. Baret, J.; Miller, O.J.; Taly, V.; El-harrak, A.; Frenz, L.; Rick, C.; Samuels, M.L.; Hutchison, J.B.; Agresti, J.J.; Link, D.R.; et al. Fluorescence-activated droplet sorting (FADS): Efficient microfluidic cell sorting based on enzymatic activity. *Lab Chip* **2009**, *9*, 1850–1858. [CrossRef] [PubMed]

31. Mazutis, L.; Gilbert, J.; Ung, W.L.; Weitz, D.A.; Griffiths, A.D.; Heyman, J.A. Single-cell analysis and sorting using droplet-based microfluidics. *Nat. Protoc.* **2013**, *8*, 54–56. [CrossRef] [PubMed]

32. El, B.; Utharala, R.; Balyasnikova, I.V.; Griffiths, A.D.; Merten, C.A. Functional single-cell hybridoma screening using droplet-based microfluidics. *Proc. Natl. Acad. Sci. USA* **2012**, *109*, 11570–11575.

33. Schmid, L.; Weitz, D.A.; Franke, T. Sorting drops and cells with acoustics: Acoustic microfluidic fluorescence-activated cell sorter. *Lab Chip* **2014**, *14*, 3710–3718. [CrossRef] [PubMed]

34. Hejazian, M.; Li, W.; Nguyen, N.-T. Lab on a chip for continuous-flow magnetic cell separation. *Lab Chip* **2015**, *15*, 959–970. [CrossRef] [PubMed]

35. Adams, J.D.; Kim, U.; Soh, H.T. Multitarget magnetic activated cell sorter. *Proc. Natl. Acad. Sci. USA* **2008**, *105*, 18165–18170. [CrossRef] [PubMed]

36. Baier, T.; Mohanty, S.; Drese, K.S.; Rampf, F.; Kim, J.; Schonfeld, F. Modelling immunomagnetic cell capture in CFD. *Microfluid. Nanofluid.* **2009**, *7*, 205–216. [CrossRef]

37. Pamme, N.; Wilhelm, C. Continuous sorting of magnetic cells via on-chip free-flow magnetophoresis. *Lab Chip* **2006**, *6*, 974–980. [CrossRef] [PubMed]

38. Zhou, Y.; Wang, Y.; Lin, Q. A microfluidic device for continuous-flow magnetically controlled capture and isolation of microparticles. *J. Microelectromech. Syst.* **2010**, *19*, 743–751. [CrossRef] [PubMed]

39. Derec, C.; Wilhelm, C.; Servais, J.; Bacri, J.C. Local control of magnetic objects in microfluidic channels. *Microfluid. Nanofluid.* **2010**, *8*, 123–130. [CrossRef]
40. Plouffe, B.D.; Lewis, L.H.; Murthy, S.K. Computational design optimization for microfluidic magnetophoresis. *Biomicrofluidics* **2011**, *5*, 013413. [CrossRef] [PubMed]
41. Issadore, D.; Shao, H.; Chung, J.; Newton, A.; Pittet, M.; Weissleder, R.; Lee, H. Self-assembled magnetic filter for highly efficient immunomagnetic separation. *Lab Chip* **2011**, *11*, 147–151. [CrossRef] [PubMed]
42. Osman, O.; Toru, S.; Dumas-Bouchiat, F.; Dempsey, N.M.; Haddour, N.; Zanini, L.F.; Buret, F.; Reyne, G.; Frenea-Robin, M. Microfluidic immunomagnetic cell separation using integrated permanent micromagnets. *Biomicrofluidics* **2013**, *7*, 054115. [CrossRef] [PubMed]
43. Song, S.-H.; Lee, H.-L.; Min, Y.H.; Jung, H.-I. Electromagnetic microfluidic cell labeling device using on-chip microelectromagnet and multi-layered channels. *Sens. Actuators B Chem.* **2009**, *141*, 210–216. [CrossRef]
44. Xia, N.; Hunt, T.P.; Mayers, B.T.; Alsberg, E.; Whitesides, G.M.; Westervelt, R.M.; Ingber, D.E. Combined microfluidic-micromagnetic separation of living cells in continuous flow. *Biomed. Microdevices* **2006**, *8*, 299–308. [CrossRef] [PubMed]
45. Kim, J.; Lee, H.H.; Steinfeld, U.; Seidel, H. Fast capturing on micromagnetic cell sorter. *IEEE Sens. J.* **2009**, *9*, 908–913. [CrossRef]
46. Wu, L.; Zhang, Y.; Palaniapan, M.; Roy, P. Wall effects in continuous microfluidic magneto-affinity cell separation. *Biotechnol. Bioeng.* **2010**, *106*, 68–75. [CrossRef] [PubMed]
47. Pekas, N.; Granger, M.; Tondra, M.; Popple, A.; Porter, M.D. Magnetic particle diverter in an integrated microfluidic format. *J. Magn. Magn. Mater.* **2005**, *293*, 584–588. [CrossRef]
48. Del Giudice, F.; Madadi, H.; Villone, M.M.; D'Avino, G.; Cusano, A.M.; Vecchione, R.; Ventre, M.; Maffettone, P.L.; Netti, P.A. Magnetophoresis "meets" viscoelasticity: Deterministic separation of magnetic particles in a modular microfluidic device. *Lab Chip* **2015**, *15*, 1912–1922. [CrossRef] [PubMed]
49. Forbes, T.P.; Forry, S.P. Microfluidic magnetophoretic separations of immunomagnetically labeled rare mammalian cells. *Lab Chip* **2012**, *12*, 1471–1479. [CrossRef] [PubMed]
50. Lee, J.; Jeong, K.J.; Hashimoto, M.; Kwon, A.H.; Rwei, A.; Shankarappa, S.A.; Tsui, J.H.; Kohane, D.S. Synthetic ligand-coated magnetic nanoparticles for micro fluidic bacterial separation from blood. *Nano Lett.* **2014**, *14*, 1–5. [CrossRef] [PubMed]
51. Kirby, D.; Siegrist, J.; Kijanka, G.; Zavattoni, L.; Sheils, O.; O'Leary, J.; Burger, R.; Ducrée, J. Centrifugo-magnetophoretic particle separation. *Microfluid. Nanofluid.* **2012**, *13*, 899–908. [CrossRef]
52. Kirby, D.; Glynn, M.; Kijanka, G.; Ducree, J. Rapid and cost-efficient enumeration of rare cancer cells from whole blood by low-loss centrifugo-magnetophoretic purification under stopped-flow conditions. *Cytom. A* **2015**, *87*, 74–80. [CrossRef] [PubMed]
53. Hoshino, K.; Huang, Y.-Y.; Lane, N.; Huebschman, M.; Uhr, J.W.; Frenkel, E.P.; Zhang, X. Microchip-based immunomagnetic detection of circulating tumor cells. *Lab Chip* **2011**, *11*, 3449–3457. [CrossRef] [PubMed]
54. Shields IV, C.W.; Livingston, C.E.; Yellen, B.B.; López, G.P.; Murdoch, D.M.; Iv, C.W.S.; Livingston, C.E.; Yellen, B.B. Magnetographic array for the capture and enumeration of single cells and cell pairs. *Biomicrofluidics* **2014**, *8*, 041101. [CrossRef] [PubMed]
55. Wang, J.; Liu, Y.; Teesalu, T.; Sugahara, K.N.; Ramana, V. Selection of phage-displayed peptides on live adherent cells in microfluidic channels. *Proc. Natl. Acad. Sci. USA* **2011**, *108*, 6909–6914. [CrossRef] [PubMed]
56. Fatanat, T.; Tabrizian, M. Adhesion based detection, sorting and enrichment of cells in microfluidic Lab-on-Chip devices. *Lab Chip* **2010**, *10*, 3043–3053.
57. Miwa, J.; Suzuki, Y.; Kasagi, N. Adhesion based cells sorter witha antibody immobilized functionalized paylene surface. In Proceedings of the IEEE 20th International Conference on Micro Electro Mechanical Systems, Hyogo, Japan, 21–25 January 2007; pp. 27–30.
58. Lo, C.Y.; Antonopoulos, A.; Dell, A.; Haslam, S.M.; Lee, T.; Neelamegham, S. The use of surface immobilization of P-selectin glycoprotein ligand-1 on mesenchymal stem cells to facilitate selectin mediated cell tethering and rolling. *Biomaterials* **2013**, *34*, 8213–8222. [CrossRef] [PubMed]
59. Gaskill, M.M.; Launiere, C.A.; Eddington, D.T. Optimization of protein immobilization in microfludic devices for circulating tumor cell capture. *J. Undergrad. Res.* **2012**, *1*, 1–5.
60. Kurkuri, M.D.; Al-ejeh, F.; Shi, Y.; Palms, D.; Prestidge, C.; Griesser, H.J.; Brown, M.P.; Thierry, B. Plasma functionalized PDMS microfluidic chips: Towards point-of-care capture of circulating tumor cells. *J. Mater. Chem.* **2011**, *21*, 8841–8848. [CrossRef]

61. Mitchell, M.J.; Castellanos, C.A.; King, M.R. Immobilized surfactant-nanotube complexes support selectin-mediated capture of viable circulating tumor cells in the absence of capture antibodies. *J. Biomed. Mater. Res. A* **2015**, *103*, 3407–3418. [CrossRef] [PubMed]

62. Zhu, J.; Nguyen, T.; Pei, R.; Stojanovic, M.; Lin, Q. Specific capture and temperature-mediated release of cells in an aptamer-based microfluidic device. *Lab Chip* **2014**, *12*, 3504–3513. [CrossRef] [PubMed]

63. Zhang, C.; Lv, X.; Han, X.; Man, Y.; Saeed, Y.; Qing, H.; Deng, Y. Analytical Methods Whole-cell based aptamer selection for selective capture of microorganisms using micro fluidic devices. *Anal. Methods* **2015**, *7*, 6339–6345. [CrossRef]

64. Jeon, S.; Hong, W.; Lee, E.S.; Cho, Y. High-purity isolation and recovery of circulating tumor cells using conducting polymer-deposited microfluidic device. *Theranostics* **2014**, *4*, 1123–1132. [CrossRef] [PubMed]

65. Bussonni, A.; Bou-matar, O.; Grandbois, M. Cell detachment and label-free cell sorting using modulated surface acoustic waves (SAW) in droplet-based microfluidics. *Lab Chip* **2014**, *14*, 3556–3563. [CrossRef] [PubMed]

66. Fatanat, T.; Li, K.; Veres, T.; Tabrizian, M. Separation of rare oligodendrocyte progenitor cells from brain using a high-throughput multilayer thermoplastic-based micro fluidic device. *Biomaterials* **2013**, *34*, 5588–5593.

67. Zheng, S.; Lin, H.; Liu, J.Q.; Balic, M.; Datar, R.; Cote, R.J.; Tai, Y.C. Membrane microfilter device for selective capture, electrolysis and genomic analysis of human circulating tumor cells. *J. Chromatogr. A* **2007**, *1162*, 154–161. [CrossRef] [PubMed]

68. Adams, D.L.; Zhua, P.; Makarovab, O.V.; Martinc, S.S.; Charpentierc, M.; Chumsric, S.; Lia, S.; Amstutzd, P.; Tangd, C.-M. The systematic study of circulating tumor cell isolation using lithographic microfilters. *RSC Adv.* **2015**, *9*, 4334–4342. [CrossRef] [PubMed]

69. Fan, X.; Jia, C.; Yang, J.; Li, G.; Mao, H.; Jin, Q. A microfluidic chip integrated with a high-density PDMS-based microfiltration membrane for rapid isolation and detection of circulating tumor cells. *Biosens. Bioelectron.* **2015**, *71*, 380–386. [CrossRef] [PubMed]

70. Xu, L.; Mao, X.; Imrali, A.; Syed, F.; Mutsvangwa, K.; Berney, D.; Cathcart, P.; Hines, J.; Shamash, J.; Lu, Y.J. Optimization and evaluation of a novel size based circulating tumor cell isolation system. *PLoS ONE* **2015**, *10*, e0138032. [CrossRef] [PubMed]

71. Chudziak, J.; Burt, D.J.; Mohan, S.; Rothwell, D.G.; Mesquita, B.; Antonello, J.; Dalby, S.; Ayub, M.; Priest, L.; Carter, L.; et al. Clinical evaluation of a novel microfluidic device for epitope-independent enrichment of circulating tumour cells in patients with small cell lung cancer. *Analyst* **2016**, *141*, 669–678. [CrossRef] [PubMed]

72. Hvichia, G.E.; Parveen, Z.; Wagner, C.; Janning, M.; Quidde, J.; Stein, A.; Muller, V.; Loges, S.; Neves, R.P.L.; Stoecklein, N.H.; et al. A novel microfluidic platform for size and deformability based separation and the subsequent molecular characterization of viable circulating tumor cells. *Int. J. Cancer* **2016**, *138*, 2894–2904. [CrossRef] [PubMed]

73. Harb, W.; Fan, A.; Tran, T.; Danila, D.C.; Keys, D.; Schwartz, M.; Ionescu-Zanetti, C. Mutational analysis of circulating tumor cells using a novel microfluidic collection device and qPCR assay. *Transl. Oncol.* **2013**, *6*, 528–538. [CrossRef] [PubMed]

74. Desitter, I.; Guerrouahen, B.S.; Benali-Furet, N.; Wechsler, J.; Janne, P.A.; Kuang, Y.; Yanagita, M.; Wang, L.; Berkowitz, J.A.; Distel, R.J.; et al. A new device for rapid isolation by size and characterization of rare circulating tumor cells. *Anticancer Res.* **2011**, *31*, 427–441. [PubMed]

75. Makarova, O.V.; Tang, C.; Amstutz, P.; Hoffbauer, M.; Williamson, T. Fabrication of high density, high-aspect-ratio polyimide nanofilters. *J. Vac. Sci. Technol. B* **2009**, *27*, 2585–2587. [CrossRef]

76. Hosokawa, M.; Hayata, T.; Fukuda, Y.; Arakaki, A.; Yoshino, T.; Tanaka, T.; Matsunaga, T. Size-selective microcavity array for rapid and efficient detection of circulating tumor cells. *Anal. Chem.* **2010**, *82*, 6629–6635. [CrossRef] [PubMed]

77. Lim, L.S.; Hu, M.; Huang, C.; Cheong, C.; Liang, T.; Looi, L. Lab on a Chip Microsieve lab-chip device for rapid enumeration and fluorescence in situ hybridization of circulating tumor cells. *Lab Chip* **2012**, *12*, 4388–4396. [CrossRef] [PubMed]

78. Chen, W.; Huang, N.; Oh, B.; Lam, R.H.W.; Fan, R.; Cornell, T.T.; Shanley, T.P.; Kurabayashi, K.; Fu, J. Surface-micromachined microfiltration membranes for efficient isolation and functional immunophenotyping of subpopulations of immune cells. *Adv. Healthc. Mater.* **2013**, *2*, 965–975. [CrossRef] [PubMed]

79. Bielefeld-Sevigny, M. AlphaLISA immunoassay platform—The "no-wash" high-throughput alternative to ELISA. *Assay Drug Dev. Technol.* **2009**, *7*, 90–92. [CrossRef] [PubMed]
80. Tang, Y.; Shi, J.; Li, S.; Wang, L.; Cayre, Y.E.; Chen, Y. Microfluidic device with integrated mciro filter of conical-shaped holes for high efficiency and high purity capture of circulating tumor cells. *Sci. Rep.* **2014**, *4*, 6052. [CrossRef] [PubMed]
81. Zheng, S.; Lin, H.K.; Lu, B.; Williams, A.; Datar, R.; Cote, R.J. 3D microfilter device for viable circulating tumor cell (CTC) enrichment from blood. *Biomed. Microdevices* **2011**, *13*, 203–213. [CrossRef] [PubMed]
82. Zhou, M.; Hao, S.; Williams, A.J.; Harouaka, R.A.; Schrand, B.; Rawal, S.; Ao, Z.; Brenneman, R.; Gilboa, E.; Lu, B.; et al. W Separable bilayer microfiltration device for viable label-free enrichment of circulating tumour cells. *Sci. Rep.* **2014**, *4*, 7392. [CrossRef] [PubMed]
83. Bhagat, A.A.S.; Kuntaegowdanahalli, S.S.; Papautsky, I. Enhanced particle filtration in straight microchannels using shear-modulated inertial migration. *Phys. Fluids* **2008**, *20*, 101702. [CrossRef]
84. Di Carlo, D.; Edd, J.F.; Irimia, D.; Tompkins, R.G.; Toner, M. Equilibrium separation and filtration of particles using differential inertial focusing. *Anal. Chem.* **2008**, *80*, 2204–2211. [CrossRef] [PubMed]
85. Hur, S.C.; Mach, A.J.; Di Carlo, D. High-throughput size-based rare cell enrichment using microscale vortices. *Biomicrofluidics* **2011**, *5*, 022206. [CrossRef] [PubMed]
86. Gregoratto, I.; McNeil, C.J.; Reeks, M.W. Micro-devices for rapid continuous separation of suspensions for use in micro-total-analysis-systems (µTAS). *Proc. SPIE* **2007**, *6465*, 646503.
87. Son, J.; Murphy, K.; Samuel, R.; Gale, B.K.; Carrell, D.T.; Hotaling, J.M. Non-motile sperm cell separation using a spiral channel. *Anal. Methods* **2015**, *7*, 8041–8047. [CrossRef]
88. Warkiani, M.E.; Guan, G.; Luan, K.B.; Lee, W.C.; Bhagat, A.A.S.; Kant Chaudhuri, P.; Tan, D.S.-W.; Lim, W.T.; Lee, S.C.; Chen, P.C.Y.; et al. Slanted spiral microfluidics for the ultra-fast, label-free isolation of circulating tumor cells. *Lab Chip* **2014**, *14*, 128–137. [CrossRef] [PubMed]
89. Johnston, I.D.; Mcdonnell, M.B.; Tan, C.K.L.; Mccluskey, D.K.; Davies, M.J.; Tracey, M.C. Dean flow focusing and separation of small microspheres within a narrow size range. *Microfluid. Nanofluid.* **2014**, *17*, 509–518. [CrossRef]
90. Burke, J.M.; Zubajlo, R.E.; Smela, E.; White, I.M. High-throughput particle separation and concentration using spiral inertial filtration. *Biomicrofluidics* **2014**, *8*, 024105. [CrossRef] [PubMed]
91. Xiang, N.; Yi, H.; Chen, K.; Sun, D.; Jiang, D.; Dai, Q.; Ni, Z. High-throughput inertial particle focusing in a curved microchannel: Insights into the flow-rate regulation mechanism and process model. *Biomicrofluidics* **2013**, *7*, 044116. [CrossRef] [PubMed]
92. Jimenez, M.; Miller, B.; Bridle, H.L. Efficient separation of small micro particles at high flowrates using spiral channels: Applicationn to waterborne pathogens. *Chem. Eng. Sci.* **2015**, in press. [CrossRef]
93. Hong, S.C.; Kang, J.S.; Lee, J.E.; Kim, S.S.; Jung, J.H. Lab on a Chip microfluidics and its application to airborne. *Lab Chip* **2015**, *15*, 1889–1897. [CrossRef] [PubMed]
94. Kuntaegowdanahalli, S.S.; Bhagat, A.S.; Papautsky, I. Inertial microfluidics for continuous particle separation in spiral microchannels. *Lab Chip* **2009**, *9*, 2973–2980. [CrossRef] [PubMed]
95. Lee, W.C.; Bhagat, A.A.S.; Huang, S.; Van Vliet, K.J.; Han, J.; Lim, C.T. High-throughput cell cycle synchronization using inertial forces in spiral microchannels. *Lab Chip* **2011**, *11*, 1359–1367. [CrossRef] [PubMed]
96. Nivedita, N.; Papautsky, I. Continuous separation of blood cells in spiral microfluidic devices. *Biomicrofluidics* **2013**, *7*, 054101. [CrossRef] [PubMed]
97. Zhou, J.; Giridhar, P.V.; Kasper, S.; Papautsky, I. Modulation of rotation-induced lift force for cell filtration in a low aspect ratio microchannel. *Biomicrofluidics* **2014**, *8*, 044112. [CrossRef] [PubMed]
98. Guan, G.; Wu, L.; Bhagat, A.A.; Li, Z.; Chen, P.C.Y.; Chao, S.; Ong, C.J.; Han, J. Spiral microchannel with rectangular and trapezoidal cross-sections for size based particle separation. *Sci. Rep.* **2013**, *3*, 1475. [CrossRef] [PubMed]
99. Khoo, B.L.; Warkiani, M.E.; Tan, D.S.W.; Bhagat, A.A.S.; Irwin, D.; Lau, D.P.; Lim, A.S.T.; Lim, K.H.; Krisna, S.S.; Lim, W.T.; et al. Clinical validation of an ultra high-throughput spiral microfluidics for the detection and enrichment of viable circulating tumor cells. *PLoS ONE* **2014**, *9*, e111296. [CrossRef] [PubMed]
100. Warkiani, M.E.; Tay, A.K.P.; Khoo, B.L.; Xiaofeng, X.; Han, J.; Lim, C.T. Malaria detection using inertial microfluidics. *Lab Chip* **2015**, *15*, 1101–1109. [CrossRef] [PubMed]

101. Warkiani, M.E.; Tay, A.K.P.; Guan, G.; Han, J. Membrane-less microfiltration using inertial microfluidics. *Sci. Rep.* **2015**, *5*, 11018. [CrossRef] [PubMed]

102. Warkiani, M.E.; Khoo, B.L.; Tan, D.S.-W.; Bhagat, A.A.S.; Lim, W.-T.; Yap, Y.S.; Lee, S.C.; Soo, R.A.; Han, J.; Lim, C.T. An ultra-high-throughput spiral microfluidic biochip for the enrichment of circulating tumor cells. *Analyst* **2014**, *139*, 3245–3255. [CrossRef] [PubMed]

103. Che, J.; Yu, V.; Dhar, M.; Renier, C.; MAtsumoto, M.; Heirich, S.; Garon, E.B.; Goldman, J.; Rao, J.; Sledge, G.W.; et al. Classification of large circulating tumor cells isolated with ultra-high throughput microfluidic vortex technology. *Oncotarget* **2016**, *7*, 12748–12760. [PubMed]

104. Lee, M.G.; Shin, J.H.; Bae, C.Y.; Choi, S.; Park, J. Label-free cancer cell separation from human whole blood using inertial micro fluidics at low shear stress. *Anal. Chem.* **2013**, *85*, 6213–6218. [CrossRef] [PubMed]

105. Wang, X.; Zhou, J.; Papautsky, I. Vortex-aided inertial microfluidic device for continuous particle separation with high size-selectivity, efficiency, and purity. *Biomicrofluidics* **2013**, *7*, 22–25. [CrossRef] [PubMed]

106. Shen, S.; Zhao, L.; Wang, Y.; Wang, J.; Xu, J.; Li, T.; Pang, L.; Wang, J. High-throughput rare cell separation from blood samples using steric hindrance and inertial microfluidics. *Lab Chip* **2014**, *14*, 2525–2538. [CrossRef] [PubMed]

107. Zhou, J.; Kasper, S.; Papautsky, I. Enhanced size-dependent trapping of particles using microvortices. *Microfluid. Nanofluid.* **2013**, *15*, 611–623. [CrossRef] [PubMed]

108. Holm, S.H.; Beech, J.P.; Barrett, M.P.; Tegenfeldt, J.O. Separation of parasites from human blood using deterministic lateral displacement. *Lab Chip* **2011**, *11*, 1326–1332. [CrossRef] [PubMed]

109. Beech, J.P.; Holm, S.H.; Adolfsson, K.; Tegenfeldt, J.O. Sorting cells by size, shape and deformability. *Lab Chip* **2012**, *12*, 1048–1051. [CrossRef] [PubMed]

110. Karabacak, N.M.; Spuhler, P.S.; Fachin, F.; Lim, E.J.; Pai, V.; Ozkumur, E.; Martel, J.M.; Kojic, N.; Smith, K.; Chen, P.; et al. Microfluidic, marker-free isolation of circulating tumor cells from blood samples. *Nat. Protoc.* **2014**, *9*, 694–710. [CrossRef] [PubMed]

111. Holmes, D.; Whyte, G.; Bailey, J.; Vergara-Irigaray, N.; Ekpenyong, A.; Guck, J.; Duke, T. Separation of blood cells with differing deformability using deterministic lateral displacement. *Interface Focus* **2014**, *4*, 20140011. [CrossRef] [PubMed]

112. Liu, Z.; Huang, F.; Du, J.; Shu, W.; Feng, H. Rapid isolation of cancer cells using microfluidic deterministic lateral displacement structure. *Biomicrofluidics* **2013**, *7*, 011801. [CrossRef] [PubMed]

113. Loutherback, K.; Silva, J.D.; Liu, L.; Wu, A.; Austin, R.H. Deterministic separation of cancer cells from blood at 10 mL/min at 10 mL/min. *AIP Adv.* **2012**, *2*, 42107. [CrossRef] [PubMed]

114. Zeming, K.K.; Ranjan, S.; Zhang, Y. Rotational separation of non-spherical bioparticles using I-shaped pillar arrays in a microfluidic device. *Nat. Commun.* **2013**, *4*, 1625–1628. [CrossRef] [PubMed]

115. Ranjan, S.; Zeming, K.K.; Jureen, R.; Fisher, D.; Zhang, Y. DLD pillar shape design for efficient separation of spherical and non-spherical bioparticles. *Lab Chip* **2014**, *14*, 4250–4262. [CrossRef] [PubMed]

116. Zeming, K.K.; Salafi, T.; Chen, C.; Zhang, Y. Asymmetrical deterministic lateral displacement gaps for dual functions of enhanced separation and throughput of red blood cells. *Sci. Rep.* **2016**, *6*, 22934. [CrossRef] [PubMed]

117. Pawell, R.S.; Taylor, R.A.; Morris, K.V.; Barber, T.J. Automating microfluidic part verification. *Microfluid. Nanofluid.* **2014**, *18*, 657–665. [CrossRef]

118. Collins, D.J.; Alan, T.; Neild, A. Particle separation using virtual deterministic lateral displacement (vDLD). *Lab Chip* **2014**, *14*, 1595–1603. [CrossRef] [PubMed]

119. Low, W.S.; Abu, W.; Wan, B. Benchtop technologies for circulating tumor cells separation based on biophysical properties. *Biomed Res. Int.* **2015**, *2015*, 239362. [CrossRef] [PubMed]

120. Haeberle, S.; Brenner, T.; Zengerle, R.; Ducrée, J. Centrifugal extraction of plasma from whole blood on a rotating disk. *Lab Chip* **2006**, *6*, 776–781. [CrossRef] [PubMed]

121. Zhang, J.; Guo, Q.; Liu, M.; Yang, J. A lab-on-CD prototype for high-speed blood separation. *J. Micromech. Microeng.* **2008**, *18*, 125025. [CrossRef]

122. Li, B.S.; Kuo, J.N. A compact disk (CD) microfluidic platform for rapid separation and mixing of blood plasma. In Proceedings of the 8th Annual IEEE International Conference on Nano/Micro Engineered and Molecular Systems, Suzhou, China, 7–10 April 2013; pp. 462–465.

123. Burger, R.; Reis, N.; da Fonseca, J.G.; Ducrée, J. Plasma extraction by centrifugo-pneumatically induced gating of flow. *J. Micromech. Microeng.* **2013**, *23*, 035035. [CrossRef]

124. Lee, A.; Park, J.; Lim, M.; Sunkara, V.; Kim, S.Y.; Kim, G.H.; Kim, M.; Cho, Y. All-in-one centrifugal micro fluidic device for size-selective circulating tumor cell isolation with high purity. *Anal. Chem.* **2014**, *86*, 11349–11356. [CrossRef] [PubMed]

125. Lu, X.; Xuan, X. Elasto-inertial pinched flow fractionation for continuous shape-based particle separation. *Anal. Chem.* **2015**, *87*, 11523–11530. [CrossRef] [PubMed]

126. Vig, A.L.; Kristensen, A. Separation enhancement in pinched flow fractionation. *Appl. Phys. Lett.* **2008**, *93*, 20–23. [CrossRef]

127. Cupelli, C.; Borchardt, T.; Steiner, T.; Paust, N.; Zengerle, R.; Santer, M. Leukocyte enrichment based on a modified pinched flow fractionation approach. *Microfluid. Nanofluid.* **2013**, *14*, 551–563. [CrossRef]

128. Song, J.; Song, M.; Kang, T.; Kim, D.; Lee, L.P. Label-free density difference amplification-based cell sorting. *Biomicrofluidics* **2014**, *8*, 064108. [CrossRef] [PubMed]

129. Morijiri, T.; Sunahiro, S.; Senaha, M.; Yamada, M.; Seki, M. Sedimentation pinched-flow fractionation for size- and density-based particle sorting in microchannels. *Microfluid. Nanofluid.* **2011**, *11*, 105–110. [CrossRef]

130. Li, P.; Mao, Z.; Peng, Z.; Zhou, L.; Chen, Y.; Huang, P.-H.; Truica, C.I.; Drabick, J.J.; El-Deiry, W.S.; Dao, M.; et al. Acoustic separation of circulating tumor cells. *Proc. Natl. Acad. Sci. USA* **2015**, *112*, 4970–4975. [CrossRef] [PubMed]

131. Petersson, F.; Lena, A.; Swa, A.; Laurell, T. Free flow acoustophoresis: Microfluidic-based mode of particle and cell separation. *Anal. Chem.* **2007**, *79*, 5117–5123. [CrossRef] [PubMed]

132. Dykes, J.; Lenshof, A.; Åstrand-Grundström, I-B.; Laurell, T.; Scheding, S. Efficient removal of platelets from peripheral blood progenitor cell products using a novel micro-chip based acoustophoretic platform. *PLoS ONE* **2011**, *6*, e23074. [CrossRef] [PubMed]

133. Yang, A.H.J.; Soh, H.T. Acoustophoretic sorting of viable mammalian cells in a microfluidic device. *Anal. Chem.* **2013**, *84*, 10756–10762. [CrossRef] [PubMed]

134. Burguillos, M.A.; Magnusson, C.; Nordin, M.; Lenshof, A.; Augustsson, P.; Hansson, M.J.; Elmér, E.; Lilja, H.; Brundin, P.; Laurell, T.; et al. Microchannel acoustophoresis does not impact survival or function of microglia, leukocytes or tumor cells. *PLoS ONE* **2013**, *8*, e64233. [CrossRef] [PubMed]

135. Ding, X.; Lin, S.-C.S.; Lapsley, M.I.; La, S.; Guo, X.; Chan, C.Y.K.; Chianga, I.-K.; Wang, L.; McCoy, J.P.; Huang, T.J. Standing surface acoustic wave (SSAW) based multichannel cell sorting. *Lab Chip* **2012**, *12*, 4228–4231. [CrossRef] [PubMed]

136. Ding, X.; Peng, Z.; Lin, S.-C.S.; Geri, M.; Li, S.; Li, P.; Chen, Y.; Dao, M.; Suresh, S.; Huang, T.J. Cell separation using tilted-angle standing surface acoustic waves. *Proc. Natl. Acad. Sci. USA* **2014**, *111*, 12992–12997. [CrossRef] [PubMed]

137. Lewpiriyawong, N.; Yang, C.; Lam, Y.C. Continuous sorting and separation of microparticles by size using AC dielectrophoresis in a PDMS microfluidic device with 3-D conducting PDMS composite electrodes. *Electrophoresis* **2010**, *31*, 2622–2631. [CrossRef] [PubMed]

138. Kang, Y.; Cetin, B.; Wu, Z.; Li, D. Continuous particle separation with localized AC-dielectrophoresis using embedded electrodes and an insulating hurdle. *Electrochim. Acta* **2009**, *54*, 1715–1720. [CrossRef]

139. Hu, X.; Bessette, P.H.; Qian, J.; Meinhart, C.D.; Daugherty, P.S.; Soh, H.T. Marker-specific sorting of rare cells using dielectrophoresis. *Proc. Natl. Acad. Sci. USA* **2005**, *102*, 15757–16761. [CrossRef] [PubMed]

140. Alshareef, M.; Metrakos, N.; Perez, E.J.; Azer, F. Separation of tumor cells with dielectrophoresis-based microfluidic chip. *Biomicrofluidics* **2013**, *7*, 011803. [CrossRef] [PubMed]

141. Gascoyne, P.R.C.; Shim, S. Isolation of circulating tumor cells by dielectrophoresis. *Cancers (Basel)* **2014**, *6*, 545–579. [CrossRef] [PubMed]

142. Piacentini, N.; Mernier, G.; Tornay, R.; Renaud, P. Separation of platelets from other blood cells in continuous-flow by dielectrophoresis field-flow-fractionation. *Biomicrofluidics* **2011**, *5*, 34122–341228. [CrossRef] [PubMed]

143. Borgatti, M.; Altomare, L.; Aruffa, M.B.; Fabbri, E.; Breveglieri, G.; Feriotto, G.; Manaresi, N.; Medoro, G.; Romani, A.; Tartagni, M.; et al. Separation of white blood cells from erythrocytes on a dielectrophoresis (DEP) based 'Lab-on-a-chip' device. *Int. J. Mol. Med.* **2005**, *15*, 913–920. [CrossRef] [PubMed]

144. Song, H.; Rosano, J.M.; Wang, Y.; Garson, C.J.; Prabhakarpandian, B.; Pant, K.; Klarmann, G.J.; Perantoni, A.; Alvarez, L.M. Continuous-flow sorting of stem cells and differentiation products based on dielectrophoresis. *Lab Chip* **2015**, *15*, 1320–1328. [CrossRef] [PubMed]

145. Holmes, D.; Green, N.G. Microdevices for Dielectrophoresis Flow-through Cell separation. *IEEE Eng. Med. Biol. Mag.* **2003**, *22*, 85–90. [CrossRef] [PubMed]

146. Chen, C.-A.; Chen, C.-H.; Ghaemmaghami, A.M.; Fan, S.-K. Separation of dendritic and T cells using electrowetting and dielectrophoresis. In Proceedings of the 7th IEEE International Conference on Nano/Micro Engineered and Molecular Systems (NEMS), Kyoto, Japan, 5–8 March 2012; pp. 83–186.

147. Li, H.; Zheng, Y.; Akin, D.; Bashir, R.; Member, S. Characterization and modeling of a microfluidic dielectrophoresis filter for biological species. *J. Microelectromech. Syst.* **2005**, *14*, 103–112. [CrossRef]

148. Moon, H.; Nam, Y. Dielectrophoretic separation of airborne microbes and dust particles using a microfluidic channel for real-time bioaerosol monitoring. *Environ. Sci. Technol.* **2009**, *43*, 5857–5863. [CrossRef] [PubMed]

149. Ling, S.H.; Lam, Y.C.; Chian, K.S. Continuous cell separation using dielectrophoresis through asymmetric and periodic microelectrode array. *Anal. Chem.* **2012**, *84*, 6463–6470. [CrossRef] [PubMed]

150. Lin, Y.; Lee, G. Optically induced flow cytometry for continuous microparticle counting and sorting. *Biosens. Bioelectron.* **2008**, *24*, 572–578. [CrossRef] [PubMed]

151. Garcia, M.M.; Ohta, A.T.; Walsh, T.J.; Vittinghof, E.; Lin, G.; Wu, M.C.; Lue, T.F. Sexual function/infertility a noninvasive, motility independent, sperm sorting method and technology to identify and retrieve individual viable nonmotile sperm for intracytoplasmic sperm injection. *J. Urol.* **2010**, *184*, 2466–2472. [CrossRef] [PubMed]

152. Huang, S.; Chen, J.; Wang, J.; Yang, C.; Wu, M. A new optically-induced dielectrophoretic (ODEP) force-based scheme for effective cell sorting. *Int. J. Electrochem. Sci.* **2012**, *7*, 12656–12667.

153. Jen, C.; Chang, H. A handheld preconcentrator for the rapid collection of cancerous cells using dielectrophoresis generated by circular microelectrodes in stepping electric fields A handheld preconcentrator for the rapid collection of cancerous cells using dielectrophoresis. *Biomicrofluidics* **2011**, *5*, 034101. [CrossRef] [PubMed]

154. Chen, G.; Huang, C.; Wu, H.; Zamay, T.N.; Zamay, A.S.; Jen, C. Isolating and concentrating rare cancerous cells in large sample volumes of blood by using dielectrophoresis and stepping electric fields. *BioChip J.* **2014**, *8*, 67–74. [CrossRef]

155. Xie, H.; Tewari, R.; Fukushima, H.; Narendra, J.; Heldt, C.; King, J.; Minerick, A.R. Development of a 3D graphene electrode dielectrophoretic device. *J. Vis. Exp.* **2014**, *88*, e51696. [CrossRef] [PubMed]

156. Abdul Razak, M.A.; Hoettges, K.F.; Fatoyinbo, H.O.; Labeed, F.H.; Hughes, M.P. Efficient dielectrophoretic cell enrichment using a dielectrophoresis-well based system. *Biomicrofluidics* **2013**, *7*, 064110. [CrossRef] [PubMed]

157. Voldman, J.; Gray, M.L.; Toner, M.; Schmidt, M.A. A microfabrication-based dynamic array cytometer. *Anal. Chem.* **2002**, *74*, 3984–3990. [CrossRef] [PubMed]

158. Martinez-Duarte, R.; Renaud, P.; Madou, M.J. A novel approach to dielectrophoresis using carbon electrodes. *Electrophoresis* **2011**, *32*, 2385–2392. [CrossRef] [PubMed]

159. Hoettges, K.F.; HUbner, Y.; Broche, L.M.; Ogin, S.L.; Kass, G.E.N.; Hughes, M.P. Dielectrophoresis-activated multiwell plate for label-free high-throughput drug assessment. *Anal. Chem.* **2008**, *80*, 2063–2068. [CrossRef] [PubMed]

160. Iliescu, C.; Yu, L.; Tay, F.E.H.; Chen, B. Bidirectional field-flow particle separation method in a dielectrophoretic chip with 3D electrodes. *Sens. Actuators B* **2007**, *129*, 1837–1840. [CrossRef]

161. Yafouz, B.; Kadri, N.A.; Ibrahim, F. Dielectrophoretic manipulation and separation of microparticles using microarray dot electrodes. *Sensors* **2014**, *14*, 6356–6369. [CrossRef] [PubMed]

162. Lapizco-Encinas, B.H.; Simmons, B.A.; Cummings, E.B.; Fintschenko, Y. Dielectrophoretic concentration and separation of live and dead bacteria in an array of insulators. *Anal. Chem.* **2004**, *25*, 1695–1704. [CrossRef] [PubMed]

163. LaLonde, A.; Gencoglu, A.; Romero-Creel, M.F.; Koppula, K.S.; Lapizco-Encinas, B.H. Effect of insulating posts geometry on particle manipulation in insulator based dielectrophoretic devices. *J. Chromatogr. A* **2014**, *1344*, 99–108. [CrossRef] [PubMed]

164. Lewpiriyawong, N.; Yang, C.; Lam, Y.C. Dielectrophoretic manipulation of particles in a modified microfluidic H filter with multi-insulating blocks. *Biomicrofluidics* **2014**, *2*, 034105. [CrossRef] [PubMed]

165. Hyoung Kang, K.; Xuan, X.; Kang, Y.; Li, D. Effects of dc-dielectrophoretic force on particle trajectories in microchannels. *J. Appl. Phys.* **2006**, *99*, 064702. [CrossRef]

166. Chen, K.P.; Pacheco, J.R.; Hayes, M.A.; Staton, S.J.R. Insulator-based dielectrophoretic separation of small particles in a sawtooth channel. *Electrophoresis* **2009**, *30*, 1441–1448. [CrossRef] [PubMed]

167. Li, M.; Li, S.; Li, W.; Wen, W.; Alici, G. Continuous particle manipulation and separation in a hurdle-combined curved microchannel using DC dielectrophoresis. *AIP Conf. Proc.* **2013**, *1542*, 1150–1153.

168. Kang, Y.; Li, D.; Kalams, S.A.; Eid, J.E. DC-Dielectrophoretic separation of biological cells by size. *Biomed. Microdevices* **2008**, *10*, 243–249. [CrossRef] [PubMed]

169. Parikesit, G.O.F.; Markesteijn, A.P.; Piciu, O.M.; Bossche, A.; Westerweel, J.; Young, I.T.; Garini, Y. Size-dependent trajectories of DNA macromolecules due to insulative dielectrophoresis in submicrometer-deep fluidic channels. *Biomicrofluidics* **2008**, *2*, 024103. [CrossRef] [PubMed]

170. Li, M.; Li, W.H.; Zhang, J.; Alici, G.; Wen, W. A review of microfabrication techniques and dielectrophoretic microdevices for particle manipulation and separation. *J. Phys. D Appl. Phys.* **2014**, *47*, 063001. [CrossRef]

171. Qian, C.; Huang, H.; Chen, L.; Li, X.; Ge, Z.; Chen, T. Dielectrophoresis for bioparticle manipulation. *Int. J. Mol. Sci.* **2014**, *15*, 18281–18309. [CrossRef] [PubMed]

172. Furlani, E.P. Magnetophoretic separation of blood cells at the microscale. *J. Phys. D Appl. Phys.* **2007**, *40*, 1313–1319. [CrossRef]

173. Nam, J.; Huang, H.; Lim, H.; Lim, C.-S.; Shin, S. Magnetic separation of malaria-infected red blood cells in various developmental stages. *Anal. Chem.* **2013**, *15*, 7316–7323. [CrossRef] [PubMed]

174. Robert, D.; Pamme, N.; Conjeaud, H.; Gazeau, F.; Iles, A.; Wilhelm, C. Cell sorting by endocytotic capacity in a microfluidic magnetophoresis device. *Lab Chip* **2011**, *11*, 1902–1910. [CrossRef] [PubMed]

175. Mizuno, M.; Yamada, M.; Mitamura, R.; Ike, K.; Toyama, K.; Seki, M. Magnetophoresis-integrated hydrodynamic filtration system for size-and surface marker-based two-dimensional cell sorting. *Anal. Chem.* **2013**, *85*, 7666–7673. [CrossRef] [PubMed]

176. Kim, U.; Soh, H.T. Simultaneous sorting of multiple bacterial targets using integrated dielectrophoretic-magnetic activated cell sorter. *Lab Chip* **2009**, *9*, 2313–2318. [CrossRef] [PubMed]

177. Huang, R.; Barber, T.A.; Schmidt, M.A.; Tompkins, R.G.; Toner, M.; Bianchi, D.W.; Flejter, W.L.; Park, M.; Services, S.; Hospital, G. A microfluidics approach for the islolation of nucleated red blood cells (NRBCs) from the peripheral blood of pregnant women. *Prenat. Diagn.* **2008**, *28*, 892–899. [CrossRef] [PubMed]

178. Huang, C.; Liu, H.; Bander, N.H.; FKirby, B.J. Enrichment of Prostate cancer cells with blood cells with a hybrid dielectrophoresis and immunocapture microfluidic system. *Biomed Microdevices* **2013**, *15*, 941–948. [CrossRef] [PubMed]

179. Gupta, A. Separation of Leukocytes. U.S. Patent 2011/0070581A1, 24 March 2011.

180. Pamme, N. Continuous flow separations in microfluidic devices. *Lab Chip* **2007**, *7*, 1644–1659. [CrossRef] [PubMed]

micromachines

MDPI

Article

Dynamical Modeling and Analysis of Viscoelastic Properties of Single Cells

Bo Wang [1,2], Wenxue Wang [1,*], Yuechao Wang [1], Bin Liu [1] and Lianqing Liu [1,*]

[1]   State Key Laboratory of Robotics, Shenyang Institute of Automation, Chinese Academy of Sciences,
      Shenyang 110016, China; wangbo@sia.cn (B.W.); ycwang@sia.cn (Y.W.); liubin@sia.cn (B.L.)
[2]   University of Chinese Academy of Sciences, Beijing 100049, China
*   Correspondence: wangwenxue@sia.cn (W.W.); lqliu@sia.cn (L.L.);
     Tel.: +86-24-2397-0215 (W.W.); +86-24-2397-0181 (L.L.)

Academic Editors: Aaron T. Ohta and Wenqi Hu
Received: 14 April 2017; Accepted: 22 May 2017; Published: 1 June 2017

Abstract: A single cell can be regarded as a complex network that contains thousands of overlapping signaling pathways. The traditional methods for describing the dynamics of this network are extremely complicated. The mechanical properties of a cell reflect the cytoskeletal structure and composition and are closely related to the cellular biological functions and physiological activities. Therefore, modeling the mechanical properties of single cells provides the basis for analyzing and controlling the cellular state. In this study, we developed a dynamical model with cellular viscoelasticity properties as the system parameters to describe the stress-relaxation phenomenon of a single cell indented by an atomic force microscope (AFM). The system order and parameters were identified and analyzed. Our results demonstrated that the parameters identified using this model represent the cellular mechanical elasticity and viscosity and can be used to classify cell types.

Keywords: dynamical modeling; mechanical properties; principal component analysis; atomic force microscopy; viscoelasticity

---

1. Introduction

Cells are the basic component units of organisms and contain important and abundant biological information. The complete genetic information of humans can be obtained from a single cell [1]. Furthermore, a cell is a complex network that contains thousands of overlapping signaling pathways. Traditional methods for describing the dynamics of this network are extremely complicated. Zhang et al. described the local dynamic behavior of a cellular signal network using 23 equations and 82 parameters [2]. Otte et al. used a 118-equation model with 177 parameters to describe the dynamics of ion channels [3]. Moreover, it is difficult to measure variations in the chemical components of the pathways of the network in a living cell. Therefore, it is difficult to analyze the global properties of cells via their underlying mechanisms, let alone to determine how to control them.

The mechanical properties of a cell reflect the structure and composition of the cytoskeleton and play a significant role in the regulation of cell physiology; therefore, they are closely related to the cell behavior, such as that in cell growth, division, differentiation, proliferation, migration, and adhesion. Recently, some studies showed that variation in the mechanical properties of cells is associated with the emergence and development of human disease [4]. Many diseases—e.g., cancer—can drastically affect the mechanical properties at the cellular level [4–6]. The development of nanotechnologies, including the atomic force microscope (AFM), magnetic and optical tweezers [7], and micropipettes [8], has enabled the measurement of the mechanical properties of single living cells. Thus, cellular mechanical information can be utilized as a label-free biomarker for cell recognition [9], early diagnosis of disease, and drug efficacy evaluation [10]. Additionally, studying the mechanical properties of single cells may provide a potential method for the detection of abnormal cells, early diagnosis of serious disease, and drug screening. Therefore, it is important to measure and quantitatively describe the mechanical properties of a single cell using a mathematical model.

The AFM has been widely used to measure and investigate the mechanical properties of living cells. To determine cellular mechanical properties using an AFM, the AFM indentation process, generally comprising three interaction phases—approach, stress-relaxation, and retraction—between the AFM tip and sample cells, is usually implemented to obtain force-indentation curves, from which the mechanical properties can be calculated using various theories and models. The Hertz model [11] is a widely-used model to describe the relationship between the force and indentation depth—i.e., the deformation of cells—and has various mathematical expressions depending on the probe shape, including pyramid [12], cone [13], and sphere [14,15]. Therefore, the Young's modulus of cells can be statistically calculated using the Hertz model for the approach phase of the force-indentation curves. However, the Hertz model has some issues that need to be addressed. For example, the Hertz model assumes that the measured materials are linearly elastic, isotropic, and lack adhesion and friction [16], which does not hold for cells. Due to the assumption of linear elasticity for cells, the Hertz model cannot explain the dynamic stress-relaxation behavior commonly observed in living cells [17]. A.H.W. Ngan et al. developed a rate-jump approach to evaluate the elastic modulus of viscoelastic materials, including polymethyl methacrylate and living cells, using AFM indentation by applying a sudden step change in the loading rate on the sample, in which the viscous response is ignored [18–21]. However, in this method, the elasticity of the sample materials is still assumed to be linear and, considering the high complexity of the viscoelastic properties of cells, the analytical solution of elasticity is not yet accurate enough to characterize the mechanics of living cells. Using a three-element standard solid model, J. R. Dutcher et al. extracted the elastic and viscous properties of the bacterial cell envelope separately from the time-dependent creep deformation curve, which resulted from a constant force [22]. Similarly, A. Yango et al. applied a linear standard solid model to calculate the elastic and viscous properties of soft materials from the creep response to the loading and unloading steps during the stress-relaxation phase of AFM indentation [23]. In these two methods, the elastic and viscous properties are decoupled from the indentation force curves with the standard linear solid model; however, both methods assume that the soft materials are a first-order system, which ignores the high orders of the complex viscoelasticity properties of living cells. Recently, Wei et al. developed a rectification approach using finite element simulation with the assumption that the cell material is viscoelastic and has acquired the viscosity and elasticity parameters that reflect the actual dynamical mechanics of cells [24]. However, this finite element simulation-based approach does not provide an analytical solution that describes the system dynamics of cells for further system analysis.

In general, the more complicated a system is, the harder it is to model based on the underlying mechanisms. Furthermore, the mechanism-based models are extremely complicated with high order and nonlinearities, especially for organisms, making the model analysis difficult. However, a system can be dynamically modelled based on the system input and output without considering its activity mechanism. In this work, a cell was considered a dynamical system, and a linear dynamical system model with cellular viscoelastic properties as system parameters was established to describe the stress-relaxation phenomenon of cells under the indentation of the AFM cantilever. The system order was determined by system identification using the Hankel matrix method, and the system parameters—i.e., the viscoelastic properties of cells—were identified using the least squares method. The viscosity and elasticity parameters were then used for cell classification. In this work, the system order of cells was not pre-assumed and was instead determined using the system identification method and experimental data. In this model, the viscosity and elasticity properties of cells are decoupled, and each is represented by multiple parameters.

2. Materials and Methods

In this section, the experimental process of obtaining the input and output curves is first described. The AFM indentation process consists of three interaction phases—approach, stress-relaxation, and retraction. Next, a general Maxwell model to describe the indentation process is given, and the process of modeling the mechanical properties of a single cell is described. After determining the general form of the cellular system model, we used the Hankel matrix method to determine the order of the general model and the least squares method to determine the parameter in the system model. Finally, we performed mechanical parameter extraction under the viscoelastic assumption to describe the mechanical properties of a single cell.

2.1. Cell Preparation

The cell lines used in this study were obtained from the Institute Pasteur of Shanghai, Chinese Academy of Sciences (Shanghai). MCF-7 cells (human breast cancer cell line), L-929 cells (mouse fibroblast cell line), Neuro-2a cells (Mus musculus brain neuroblastoma cell line), and HEK-293 cells (human embryonic kidney cell line) were cultured in RPMI-1640 (Thermo Scientific HyClone, Logan, UT, USA) containing 10% fetal bovine serum and 1% penicillin-streptomycin solution at 37 °C (5% CO_2). These four types of cells were cultured in Petri dishes. The diameter of the Petri dishes we used was 60 mm, and the cell concentration was about 1.3×10^6 cm^{-2}. The cells were cultured for 24 h before experiments. The same experiments were performed with different batches of cells. All the AFM experiments in this study were performed in culture medium. The experiments were conducted at room temperature. In this study, twenty cells of each type were selected, and two complete indentation processes were implemented independently for each cell. Therefore, 160 force curves were recorded, and the system order identification and parameter estimation process was performed for all cells using the experimental data.

2.2. Indentation Process

To determine the mechanical properties of cells using the stress-relaxation curve with an AFM, an indentation process was implemented, and the schematic of the AFM indentation is shown in Figure 1a,b. The entire process consists of three interaction phases—i.e., approach, stress-relaxation, and retraction. During the approaching phase, the AFM probe tip is pressed on the sample cell, causing a fast cell deformation and a rapidly increasing force on the cantilever. During the stress relaxation phase, the piezoelectric (PZT) actuator is kept at a constant depth, but the cell continues to deform and cause gradual decreases in the cantilever deflection. During the retraction phase, as the AFM probe tip is retracted, the cell recovers and the cell deformation rapidly decreases. Notably, the distance that the cell is indented is not equal to the distance $u(t)$, that the PZT moves, and the difference between these two distances is the cantilever deflection. Furthermore, during the stress-relaxation phase, the deformation rate of cells becomes slower, and the cell gradually approaches constant deformation; therefore, the force between the tip and cell become constant. In this study, we used a Bioscope Catalyst AFM (Bruker, Billerica, MA, USA) and an inverted microscope (Nikon, Tokyo, Japan). The type of AFM probe used in this study was MLCT (Bruker), and the nominal spring constant of the cantilever used was 0.01 N/m. We used the thermal tune to calculate the spring constant of the cantilever. And the actual value of the spring constant was about 0.2960 N/m. The width of the cantilever was about 0.59–0.61 µm, and the aspect ratio of the AFM tip was about 3.78. In this study, we used the same cantilever to probe all the cells in the experiment. The movement speed of the AFM PZT in the indentation experiments was 4 µm/s during both the approach and retraction phases. The stress-relaxation time was 6 s. In the experiment, the maximum distance of cantilever displacement was 1 µm. As mentioned above, the indentation depth on the cells was not equal to the distance that the PZT moved. The indentation depth of each cell was about 0.6–0.7 µm. The thicknesses of the cells were about 4–7 μm. For all of the AFM experiments we performed in this study, we measured the stress-relaxation curves of the cells in the nuclear region.

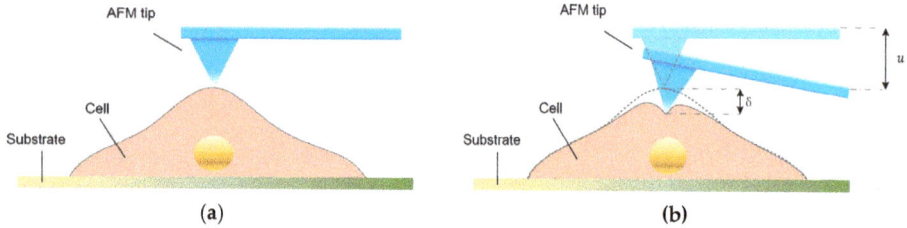

Figure 1. Illustration of the atomic force microscope (AFM) indentation experiment and the experimental curve from one entire indentation process. (**a,b**) Schematic diagram of indentation applied using an AFM probe tip on a single cell.

2.3. Dynamical Modeling of the Viscoelastic Properties of a Single Cell

In this study, system sciences were used to model the dynamical mechanical behavior of a cell with its viscoelastic properties as the system parameters, based on the input (stimuli) and corresponding output (responses) instead of the interior structure of the cells. In this approach, as shown in Figure 2a, a single cell is considered as a general system and generates an output response under a certain input stimulus. The state variable *x* describes the internal dynamics of cell systems and constitutes the system output response *y* under the input stimuli *u*. During the AFM indentation process, as the AFM PZT moves down, the probe tip on the cantilever presses on the cell and causes the cell to deform. On the other hand, the indentation depth is less than the PZT movement distance, which leads to the deflection of cantilever. The cantilever deflection reflects the interaction force between the probe tip and the sample cell, and therefore, the interaction force can be measured from the cantilever deflection with the coefficients of the cantilever. Therefore, in this study, the AFM PZT z-position during the indentation is used as the input signal $u(t)$, and the measured force is used as the output response $y(t)$ of the cell system. In this study, the dynamic behaviors of a cell during the indentation process is modelled using a system approach without pre-assumption of the system order. The system order is determined with a system science approach. One of such approaches is the Hankel matrix method, in which a Hankel matrix is constructed from the impulse response series. The signals during the stress-relaxation phase can be regarded as the step response to the constant input (the PZT distance) to the system, and we can obtain the impulse response series by calculating the difference between every two adjacent points in the step response sequence. Therefore, for convenience, only the signals during the stress-relaxation phase are used for modeling and further analysis. The output force curve during the stress-relaxation phase showed the tendency for an exponential decay under a constant input for the PZT z-position. Therefore, a general Maxwell model for viscoelastic materials containing a spring and *n* parallel spring-damping paths, as shown in Figure 2b, was used to model the cell dynamics of the cell system. Mathematically, the cell system can then be written using a state-space equation as follows:

$$\begin{cases} \dot{x}_i = -\frac{k_i}{b_i}x_i + \frac{k_i}{b_i}u \,, \ i = 1, 2, \ldots, n \\ y = -\sum_{i=1}^{n} k_i x_i + (k_0 + \sum_{i=1}^{n} k_i)u \end{cases} \tag{1}$$

where *u* and *y* are the system input and output, respectively, the state variable x_i represents the movement distance of the point between the spring and damper in the *i*th path, and k_i and b_i are the elastic and viscous parameters of the corresponding springs and dampers, respectively. The states x_i were closely related to the cell deformation.

This model illustrates that the elasticity and viscosity of cells can be represented by the multiple parameters k_i and b_i. Next, with the data collected in the indentation experiments, the order n of the cell system and parameters can be determined by system identification methods.

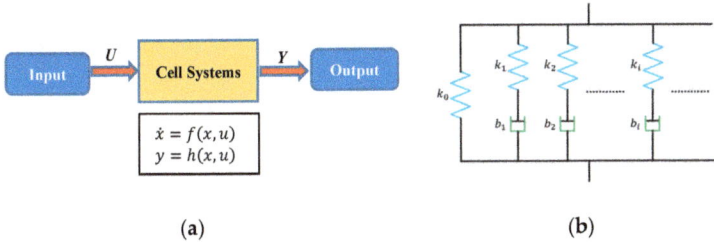

(a)   (b)

Figure 2. (**a**) Schematic diagram showing the system sciences point of view of cell dynamics; (**b**) An nth-order general Maxwell model of viscoelastic materials.

2.4. Order and Parameters Identification

The dynamical deformation behavior of a cell subjected to a constant indentation depth has been modelled by a linear dynamical model based on the structure of the general Maxwell model. In this section, the order and parameters of the cell system need to be determined from the input and output data. In this study, we used the Hankel matrix method to determine the order of the linear system. For linear systems, the Hankel matrix method is a classical approach for determining the order [25]. In this method, Hankel matrices are built from the impulse response sequence of the system, and the order of the system is actually the rank of the Hankel matrices. The criterion for identifying a system order using Hankel matrices is described in the following lemma.

Lemma 1. *Let* $\{g(i)|i = 1, 2, \ldots, L\}$ *be the impulse response sequence of a linear system. Hankel matrices can be constructed as follows:*

$$H(l,k) = \begin{bmatrix} g(k) & g(k+1) & \cdots & g(k+l-1) \\ g(k+1) & g(k+2) & & g(k+l) \\ \vdots & & \ddots & \vdots \\ g(k+l-1) & g(k+l) & \cdots & g(k+2l-1) \end{bmatrix}$$

where l determines the dimension of the Hankel matrix, and k is any integer between 1 and $L - 2l + 2$. The order of the system is equal to the rank n_0 of the Hankel matrices if

$$\text{rank}[H(l,k)] = n_0, \; for \; all \; l \geq n_0, \; \forall k \tag{2}$$

This criterion works perfectly for noise-free data. In general, the impulse response sequence includes noise, so the rank of the Hankel matrix may be not equal to n_0 exactly, even when $l \geq n_0$. Therefore, an equivalent criterion using the average ratio of the determinant of Hankel matrix D_l is used to evaluate the singularity of the Hankel matrices and the order of the dynamical systems, where

$$D_l = \frac{\frac{1}{L-2l+2}\sum_{k=1}^{L-2l+2} \det[H(l,k)]}{\frac{1}{L-2l}\sum_{k=1}^{L-2l} \det[H(l,k)]} \tag{3}$$

in which l is the dimension of the Hankel matrices and is not equal to 1. The determinant D_l grows as l increases if $l < n_0$ and then decays for $l > n_0$, and reaches a maximum at $l = n_0$. Therefore, the value of l at which D_l reaches the maximum value can be considered to be the order of the system. In this study, we used D_l to evaluate the order of the cell system.

In practice, the impulse response sequence of a dynamical system can be obtained by calculating the difference between every two adjacent points in the step response sequence of the system, i.e., $g(i) = y(i+1) - y(i)$, $i = 1, 2, \ldots, L$, where $g(i)$ is the ith element in the impulse response sequence, and $y(i)$ is the ith element in the step response sequence. In our study, the input stimulus is a constant indentation depth during the stress-relaxation phase of the indentation process. Therefore, the output of the cell system, the interaction force between the probe tip and the cell, is actually the step response. Hence, we can obtain the impulse response sequence from the recorded force curves and construct the Hankel matrices.

Once the order of the dynamic system of a single cell is determined, the parameters of the system can be easily determined by the least squares method. In this method, the parameters are chosen to minimize the error between the system model output and experimental data, i.e.;

$$\theta^* = \underset{\theta \in \Theta}{argmin} \|y_\theta(t) - \overline{y}(t)\|_2 \tag{4}$$

where $y_\theta(t)$ is the model output with the parameter $\theta = [k_0, k_1, b_1, k_2, b_2]$ in Equation (1), Θ is the corresponding parameter space of θ, $\overline{y}(t)$ is the actual output of a single cell measured by AFM in the experiments, and the parameter θ^* represents the system parameter minimizing the error. The system identification toolbox of MATLAB (R2016b, MathWorks, Natick, MA, USA) is used in this study to identify the system parameters using the least squares method.

Now we have introduced approaches for identifying the system order and estimating the system parameters of dynamical cell systems. Given the system order and parameters, the corresponding dynamic equations can describe the dynamical behavior during the stress-relaxation process and the mechanical properties of a single cell completely under the condition of viscoelastic assumptions.

3. Results and Discussion

To validate the dynamical model for cell deformation behaviors under a constant indentation depth, we performed indentation experiments using four types of cells, namely MCF-7, HEK-293, L-929, and Neuro-2a cells, and the corresponding stress-relaxation force curves were collected for system order and parameter analysis. The cells responded with different dynamic behaviors at different stages of the indentation experiment, as shown in Figure 3a. In this study, each indentation experiment lasted about 8 s, including 1 s for the approaching phase, 6 s for the stress-relaxation phase, and 1 s for the retraction phase. During the approach phase, the AFM probe tip was pressed onto the sample cells and moved down along with the constant movement of the PZT (blue curve in Figure 3a), resulting in a fast cell deformation (green curve in Figure 3a) and a rapidly increasing force on the AFM cantilever (red curve in Figure 3a). During the stress-relaxation phase, the PZT was kept at a constant depth (z-position) which was equal to 1 μm, however, the cell continued to be further deformed due to its viscosity and therefore the cantilever deflection and the interaction force between the tip and cell gradually decreased. During the retraction phase, as the AFM probe tip was retracted, the cell recovered, and therefore the cell deformation and the interaction force on the cantilever rapidly decreased. The relaxation time of a cell during the stress-relaxation phase in the indentation process depends on its mechanical properties, which were characterized by the time constants, that is to say, by the ratio of the elasticity parameters to the viscosity parameters, as shown in Equation (5). Therefore, for different types of cells, the relaxation times are different. In general, the stress-relaxation time needs to last until the stress-relaxation curve remains steady after a decline. Darling et al. showed that the stress-relaxation curves of the chondrosarcoma cells remained steady after 20 s [17]. Moreover, the response time of cells is also related to the type of stimulus signal [16,26–28]. In our study, for each type of cell, the stress-relaxation curve remained steady after 4 s. The indentation depth was about 0.6–0.7 μm, and the maximum interaction force between the tip and the cell was less than 5 nN. In this study, only the force curves during the stress-relaxation phase were used to validate the dynamical model and to identify the system parameters. To reduce the noisy effect in the measurement, a low-pass filter with a 10-Hz cut-off frequency was used to smooth the force curves for further analysis, as shown in Figure 3b.

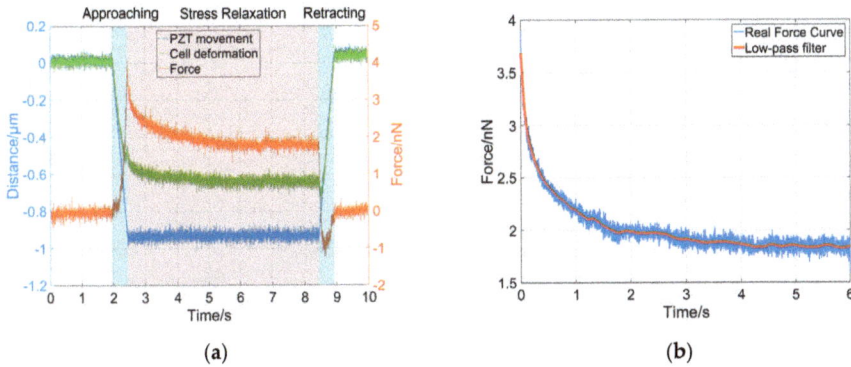

Figure 3. Illustration of the experimental curve from one entire indentation process. (**a**) The entire indentation process showing the changes in the z-position of piezoelectric (PZT) (blue), cell deformation (green), and interaction force (red) that the AFM measures during the three interaction phases; (**b**) Original force curve (blue) in the stress-relaxation phase and smoothed curve using a low-pass filter.

As shown in Figure 4a, the determinant D_l for an MCF-7 cell reaches the maximum at $l = 2$; therefore, the dynamical system for this cell is determined to be a second-order system with five parameters, including three elasticity property parameters and two viscosity property parameters. Additionally, the output solution describing the cell dynamics can be written as follows:

$$y(t) = k_0 u(t) + u(t)k_1 e^{-\frac{k_1}{b_1}t} + u(t)k_2 e^{-\frac{k_2}{b_2}t} \tag{5}$$

Equation (5) indicates that the output of the system contains three components. If the system input $u(t)$ is a constant, then the system output $y(t)$ consists of a constant component and two exponential decay components. The five parameters were estimated using the least squares method, and then the system output could be obtained accordingly. As shown in Figure 4b, the system model output (red curve) of the deformation dynamics for the MCF-7 cell fits the experimental data (blue curve) very well, and two exponential decay components were plotted, indicating that the cell deformation dynamics is mainly dominated by a fast response at the very beginning of the stress-relaxation phase and then by a slow response for the remaining time.

For all the cells, the dynamical system was determined to be second order. The system parameters were averaged for each type of cell, as indicated in Table 1. The vectors of these five parameters represent the elasticity (k_i) and viscosity (b_i) properties of a single cell; therefore, the vectors can be used to classify the cell types. Before classifying the cell types, we first conducted dimension reduction for the viscosity parameter vector to visualize the variations in the viscoelasticity parameters of cells in 2D spaces. In this study, the principal component analysis (PCA) method was used to reduce the dimension of the parameter vector. The main idea of PCA is to calculate the eigenvalue of the covariance matrix of the sample matrix. The first principle component has the largest eigenvalue, so it has the largest distinction degree. We calculated the sample matrix mentioned above, with the dimension of 160 × 5, and found that the eigenvalues of the first two principle components were larger than one, and the total contribution of the first three principle components was 96.35%. This finding indicates that the first three principle components include approximately 96.35% of the information of all five parameters (Table 2). The first three principle components of the four different types of cells are shown in Figure 5. Figure 5 shows that different types of cells present different clustering patterns; MCF-7 and Neuro-2a cells have more concentrated cluster patterns than the other two types of cells. This phenomenon may be attributed to the morphology variances of these four types of cells. As shown in Figure 6, MCF-7 and Neuro-2a cells each showed great morphology similarity, but HEK-293 and L-929 cells showed substantial variations in their shape and size. Multiple elasticity and viscosity parameters can be used to classify cell types.

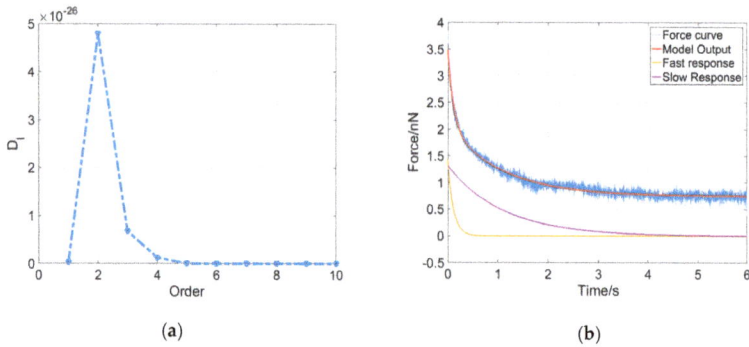

(a)   (b)

Figure 4. Determination of the system order and parameters of an MCF-7 cell. (a) D_l series of an MCF-7 cell. As l increases, D_l reaches the maximum at $l = 2$ and then decays to 0; (b) The model output of the cell system (red) of second order with estimated parameters fits the experimental force curve (blue) very well. The two exponential decay components represent the fast response (yellow) and slow response (purple) of the characteristics and the system dynamics of the MCF-7 cell.

In this study, we used a backpropagation (BP) neural network to classify the four types of cells. The neural network has two hidden layers. The first hidden layer has 20 nodes and the second one has 40 nodes. The input of the neural network is the five parameters, k_0, k_1, b_1, k_2, b_2, and the output of the neural network is the type of cell, represented by 1 for MCF-7, 2 for L-929, 3 for Neuro-2a, and 4 for HEK-293. The training data set contains 120 five-parameter tuples, 30 for each type of cells. The test dataset contains 40 five-parameter tuples, 10 for each type of cell. The algorithm was implemented with MATLAB. The classification result using the BP neural network is shown in Figure 7. Among all 40 test data for the four cell types, only two points of Neuro-2a were misclassified. The success rate of the classification is 95%; this result further proves the validity of the model for characterizing the mechanical properties of cells.

Table 1. Average parameters of cell four types.

Cellular Types	Elastic Parameters (N/m)			Viscosity Parameters (N·s/m)	
	k_0	k_1	k_2	b_1	b_2
MCF-7	2.984 ± 0.742	1.110 ± 0.372	1.385 ± 0.114	0.118 ± 0.306	3.108 ± 1.618
Neuro-2a	2.425 ± 0.589	1.519 ± 0.286	1.259 ± 0.121	0.242 ± 0.296	1.934 ± 0.891
Hek-293	1.099 ± 0.496	0.472 ± 0.203	0.609 ± 0.090	0.158 ± 0.244	1.956 ± 0.796
L-929	0.534 ± 0.096	0.195 ± 0.067	0.235 ± 0.009	0.0124 ± 0.049	0.374 ± 0.150

Table 2. Results of principle component analysis of cellular system parameters.

Principle Component	Eigenvalue	Difference Value	Contribution on Rate (%)	Total (%)
1st	3.3158	2.2298	66.3153	66.3153
2nd	1.0860	0.6685	21.7197	88.0351
3rd	0.4175	0.2631	8.3498	96.3849
4th	0.1544	0.1281	3.0884	99.4733
5th	0.0263	-	0.5267	100

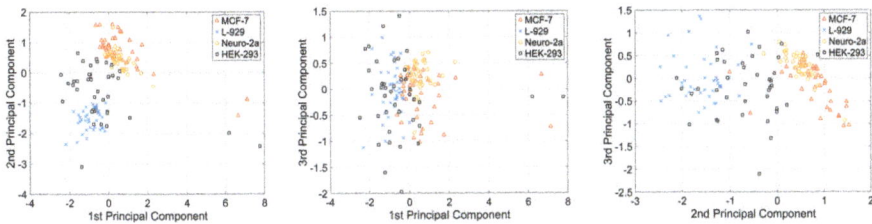

Figure 5. Principle components for four types of cells. These three plots show the first component vs. the second component, the first component vs. the third component, and the second component vs. the third component.

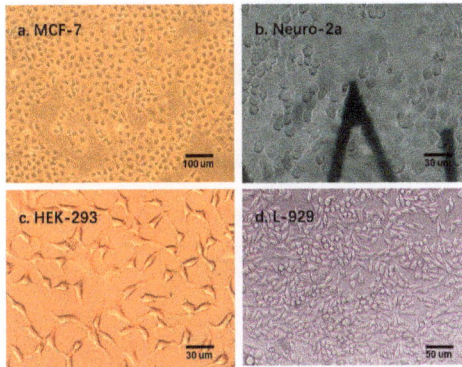

Figure 6. Morphology of four types of cells. (**a**) MCF-7 cells, (**b**) Neuro-2a cells, (**c**) HEK-293 cells and (**d**) L-929 cells. MCF-7 and Neuro-2a cells each showed less variance in morphology than HEK-293 and L-929 cells. The magnification is 10, 40, 40, and 20, respectively.

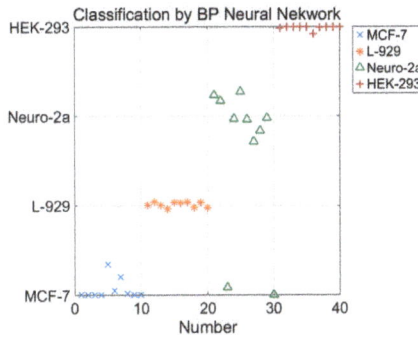

Figure 7. Classification result of the four types of cells by the backpropagation (BP) neural network. In the testing set, each type of cell has 10 test data. Among all 40 test data for the four cell types, only two points of Neuro-2a were misclassified. The success rate of the classification is 95%.

4. Conclusions

In this study, using system science approaches, we presented a dynamical model based on the structure of a general Maxwell system to describe the cell deformation dynamics under conditions of a constant indentation depth during the stress-relaxation phase of the indentation process. The system order and parameters were identified using the Hankel matrix method and the least squares method, respectively. The proposed model was validated by AFM indentation experiments with different types of cells. Four types of cells with 20 cells of each type were evaluated to collect the stress-relaxation data to validate the proposed dynamical model; the dynamical system for all the cells used in this study was determined to be second order. Therefore, the system model contained three elasticity parameters and two viscosity parameters, which represent the elasticity and viscosity characteristics, respectively, of cell dynamics. In other words, with the proposed model, the nonlinear elasticity and viscosity properties of a single cell can be decoupled, and the nonlinearity characteristics of each property can be described by multiple parameters in a linear system. PCA of the five viscoelasticity parameters revealed different clustering patterns for the four types of cells, implying that the elasticity and viscosity parameters can be used to recognize the cell type. Future work will be conducted to investigate the effects of drugs on the elasticity and viscosity parameters of cells using the proposed model.

Acknowledgments: This work was supported by the National Natural Science Foundation of China (Grant Nos. 61433017, 61673372, 61327014 and 61522312) and the CAS/SAFEA International Partnership Program for Creative Research Teams.

Author Contributions: B.W. and W.W. conceived and designed the experiments; B.W. and B.L. performed the experiments and analyzed the data; W.W., Y.W. and L.L. contributed reagents and materials; B.W. and W.W. wrote the paper. All authors read and approved the final manuscript.

Conflicts of Interest: The authors declare no conflict of interest.

References

1. Zong, C.; Lu, S.; Chapman, A.R.; Xie, X.S. Genome-wide detection of single-nucleotide and copy-number variations of a single human cell. *Science* **2012**, *338*, 1622–1626. [CrossRef] [PubMed]
2. Zhang, X.P.; Liu, F.; Wang, W. Two-phase dynamics of p53 in the DNA damage response. *Proc. Natl. Acad. Sci. USA* **2011**, *108*, 8990–8995. [CrossRef] [PubMed]
3. Otte, S.; Berg, S.; Luther, S.; Parlitz, U. Bifurcations, chaos and sensitivity to parameter variations in the sato cardiac cell model. *Commun. Nonlinear Sci. Numer. Simul.* **2016**, *37*, 265–281. [CrossRef]
4. Plodinec, M.; Loparic, M.; Monnier, C.A.; Obermann, E.C.; Zanetti-Dallenbach, R.; Oertle, P.; Hyotyla, J.T.; Aebi, U.; Bentires-Alj, M.; Lim, R.Y.; et al. The nanomechanical signature of breast cancer. *Nat. Nanotechnol.* **2012**, *7*, 757–765. [CrossRef] [PubMed]
5. Yang, W.; Yu, H.; Li, G.; Wang, B.; Wang, Y.; Liu, L. Regulation of breast cancer cell behaviours by the physical microenvironment constructed via projection microstereolithography. *Biomater. Sci.* **2016**, *4*, 863–870. [CrossRef] [PubMed]
6. Yang, W.; Yu, H.; Li, G.; Wang, Y.; Liu, L. Facile modulation of cell adhesion to a poly (ethylene glycol) diacrylate film with incorporation of polystyrene nano-spheres. *Biomed. Microdevices* **2016**, *18*, 1–7. [CrossRef] [PubMed]
7. Li, X.; Yang, H.; Wang, J.; Sun, D. Design of a robust unified controller for cell manipulation with a robot-aided optical tweezers system. *Automatica* **2015**, *55*, 279–286. [CrossRef]
8. Guilak, F.; Tedrow, J.R.; Burgkart, R. Viscoelastic properties of the cell nucleus. *Biochem. Biophys. Res. Commun.* **2000**, *269*, 781–786. [CrossRef] [PubMed]
9. Shimizu, Y.; Kihara, T.; Haghparast, S.M.; Yuba, S.; Miyake, J. Simple display system of mechanical properties of cells and their dispersion. *PLoS ONE* **2012**, *7*, e34305. [CrossRef] [PubMed]
10. Lee, G.Y.; Lim, C.T. Biomechanics approaches to studying human diseases. *Trends Biotechnol.* **2007**, *25*, 111–118. [CrossRef] [PubMed]
11. Hertz, H. Ber die berührung fester elastischer körper. *J. Reine Angew. Math.* **1881**, *92*, 156–171.
12. Rico, F.; Roca-Cusachs, P.; Gavara, N.; Farre, R.; Rotger, M.; Navajas, D. Probing mechanical properties of living cells by atomic force microscopy with blunted pyramidal cantilever tips. *Phys. Rev. E Stat. Nonlinear Soft Matter Phys.* **2005**, *72*, 021914. [CrossRef] [PubMed]
13. Laurent, V.M.; Kasas, S.; Yersin, A.; Schaffer, T.E.; Catsicas, S.; Dietler, G.; Verkhovsky, A.B.; Meister, J.J. Gradient of rigidity in the lamellipodia of migrating cells revealed by atomic force microscopy. *Biophys. J.* **2005**, *89*, 667–675. [CrossRef] [PubMed]
14. Johnson, K.L. *Contact Mechanics*; Cambridge University Press: Cambridge, MA, USA, 1985.

15. Dimitriadis, E.K.; Horkay, F.; Maresca, J.; Kachar, B.; Chadwick, R.S. Determination of elastic moduli of thin layers of soft material using the atomic force microscope. *Biophys. J.* **2002**, *82*, 2798–2810. [CrossRef]
16. Mahaffy, R.E.; Park, S.; Gerde, E.; Kas, J.; Shih, C.K. Quantitative analysis of the viscoelastic properties of thin regions of fibroblasts using atomic force microscopy. *Biophys. J.* **2004**, *86*, 1777–1793. [CrossRef]
17. Darling, E.M.; Zauscher, S.; Block, J.A.; Guilak, F. A thin-layer model for viscoelastic, stress-relaxation testing of cells using atomic force microscopy: Do cell properties reflect metastatic potential? *Biophys. J.* **2007**, *92*, 1784–1791. [CrossRef] [PubMed]
18. Tang, B.; Ngan, A.H.W. Nanoindentation using an atomic force microscope. *Philos. Mag.* **2011**, *91*, 1329–1338. [CrossRef]
19. Tang, B.; Ngan, A.H.W. A rate-jump method for characterization of soft tissues using nanoindentation techniques. *Soft Matter* **2012**, *8*, 5974–5979. [CrossRef]
20. Tang, B.; Ngan, A.H.W.; Pethica, J.B. A method to quantitatively measure the elastic modulus of materials in nanometer scale using atomic force microscopy. *Nanotechnology* **2008**, *19*, 495713. [CrossRef] [PubMed]
21. Zhou, Z.L.; Ngan, A.H.W.; Tang, B.; Wang, A.X. Reliable measurement of elastic modulus of cells by nanoindentation in an atomic force microscope. *J. Mech. Behav. Biomed. Mater.* **2012**, *8*, 134–142. [CrossRef] [PubMed]
22. Vadillo-Rodríguez, V.; Dutcher, J.R. Viscoelasticity of the bacterial cell envelope. *Soft Matter* **2011**, *7*, 4101–4110. [CrossRef]
23. Yango, A.; Schape, J.; Rianna, C.; Doschke, H.; Radmacher, M. Measuring the viscoelastic creep of soft samples by step response afm. *Soft Matter* **2016**, *12*, 8297–8306. [CrossRef] [PubMed]
24. Wei, F.; Yang, H.; Liu, L.; Li, G. A novel approach for extracting viscoelastic parameters of living cells through combination of inverse finite element simulation and atomic force microscopy. *Comput. Methods Biomech. Biomed. Eng.* **2017**, *20*, 373–384. [CrossRef] [PubMed]
25. Yin, H.; Zhu, Z.; Ding, F. Model order determination using the hankel matrix of impulse responses. *Appl. Math. Lett.* **2011**, *24*, 797–802. [CrossRef]
26. Mahaffy, R.E.; Shih, C.K.; MacKintosh, F.C.; Käs, J. Scanning probe-based frequency-dependent microrheology of polymer gels and biological cells. *Phys. Rev. Lett.* **2000**, *85*, 880–883. [CrossRef] [PubMed]
27. Raman, A.; Trigueros, S.; Cartagena, A.; Stevenson, A.P.Z.; Susilo, M.; Nauman, E.; Contera, S.A. Mapping nanomechanical properties of live cells using multi-harmonic atomic force microscopy. *Nat. Nanotechnol.* **2011**, *6*, 809–814. [CrossRef] [PubMed]
28. Cartagena, A.; Raman, A. Local viscoelastic properties of live cells investigated using dynamic and quasi-static atomic force microscopy methods. *Biophys. J.* **2014**, *106*, 1033–1043. [CrossRef] [PubMed]

micromachines

MDPI

Article

Localized Single-Cell Lysis and Manipulation Using Optothermally-Induced Bubbles

Qihui Fan [1,2], Wenqi Hu [2] and Aaron T. Ohta [2,*]

[1] College of Information Science and Technology, Beijing University of Chemical Technology,
     Beijing 100029, China; fanqihui@hawaii.edu
[2] Department of Electrical Engineering, University of Hawaii at Manoa, Honolulu, HI 96822, USA;
     wenqi@is.mpg.de
*   Correspondence: aohta@hawaii.edu; Tel.: +1-808-956-8196; Fax: +1-808-956-3427

Academic Editor: Fan-Gang Tseng
Received: 3 March 2017; Accepted: 7 April 2017; Published: 11 April 2017

Abstract: Localized single cells can be lysed precisely and selectively using microbubbles optothermally generated by microsecond laser pulses. The shear stress from the microstreaming surrounding laser-induced microbubbles and direct contact with the surface of expanding bubbles cause the rupture of targeted cell membranes. High-resolution single-cell lysis is demonstrated: cells adjacent to targeted cells are not lysed. It is also shown that only a portion of the cell membrane can be punctured using this method. Both suspension and adherent cell types can be lysed in this system, and cell manipulation can be integrated for cell–cell interaction studies.

Keywords: cell poration; cell lysis; optothermal; cell manipulation

---

1. Introduction

Intracellular components provide crucial information for biomedical research. However, the lysis of cells in bulk blends the intracellular components of all the cells, possibly resulting in misleading data due to averaging [1]. In contrast, single-cell analysis can enable the precise study of intracellular processes, like a cell's response to stimuli, such as soluble factors, cell–cell contact, and mechanical forces [1,2]. Part of many single-cell analysis procedures is the controlled lysis of specific single cells. The single-cell lysis procedure directly affects the downstream analysis, and depends upon parameters such as the cell lysing speed, throughput, selectivity, and compatibility with other processes.

There are various single-cell lysis techniques, each with its own advantages for downstream analysis. Chemical lysis uses solutions containing surfactants to solubilize lipids and proteins in the cell membrane, creating pores and eventually lysing the whole membrane. Microfluidic devices have enabled the mixing of the surfactant solutions and single cells to enable single-cell chemical lysis [2,3]. However, chemical lysis takes 2–30 s depending on the lysis solution, and the chemical reagents in the solution that need to be removed before subsequent cell analysis [2]. Electrical lysis is another common single-cell lysis technique, in which the cell membrane is broken down by an electric field of sufficient strength. Single-cell lysis can be realized with microelectrodes in a serial process [2,4], with pre-patterned microfluidic chips in a parallel process [5–7], or with virtual microelectrodes induced by light [8,9]. However, the working medium in electrical lysis is usually a sucrose solution with a relatively low electrical conductivity to decrease joule heating and bubble formation [6,10]; this is a significantly different environment from cell culture media. Cell lysis by sonication uses ultrasonic waves to generate sufficient pressure to rupture cell membranes [11], while thermal lysis destroys the cell membrane using heat [12]. These two methods have some limitations for single-cell lysis, since it is difficult to focus ultrasonic or thermal energy to a micrometer-scale localized region, and the induced temperatures may damage proteins [2,10,11]. Mechanical lysis directly uses mechanical force to rupture the cell membrane, as in the 'nanoknives' used in microfluidic devices [13,14]. Mechanical lysis can minimize the damage to proteins as compared to chemical lysis, thermal heating, and electrical lysis, although the proteins can be captured in the cell debris, increasing the difficulty of extraction [2,10,13].

Laser energy is promising for lysing localized single cells. Laser pulses are focused near the targeted cells in the cell medium to produce a cavitation bubble [2,10]. The rapid expansion or collapse of the induced cavitation bubble induces a hydrodynamic force that can lyse the cell membrane [15–19]. Laser lysis can be simply integrated into microfluidic chips, since it only requires an optical path to the targeted cell regions, and there is no need for complex channels or microelectrodes. Laser lysis can be a rapid, localized, and parallel process for targeted cells. Laser lysis has been demonstrated on single cells using various microfluidic structures, like a polydimethylsiloxane (PDMS) microchannel to confine the shock wave from laser-induced microbubble, or small holding structures to hold single cells under lysis in position [10,19,20]. These microfluidic structures facilitate the focusing of hydrodynamic force on the cells for lysis, but it limits the locations on the device where cell lysis can occur.

This paper reports on a more precise and flexible single-cell laser-lysis method, which does not need complex microfluidic structures, and can also be combined with on-chip cell manipulation. This single-cell laser lysis system utilizes microsecond laser pulses to generate size-oscillating vapor microbubbles in a biocompatible medium via an optically-absorbent substrate. During the laser on and off cycles, the microbubbles rapidly expand and collapse repeatedly, thus inducing microstreaming and the accompanying hydrodynamic forces around the microbubble. The induced hydrodynamic force can lyse a targeted single cell positioned above the microbubble. The position of the microbubble can be controlled by manipulating the laser focus spot on the substrate, so each single cell can be selectively and precisely lysed with high resolution. This system can also be used for single-cell manipulation by changing the laser pulse width and the position of laser focus [21]. The single-cell manipulation and patterning technique [22] is helpful for studying intercellular [23] and cell-environment [24] interactions. The integration of cell-patterning with a single-cell lysis system is an important feature for single-cell analysis.

2. Setup and Mechanism

2.1. Setup for Single-Cell Lysis System

The single-cell lysis system consists of a microsecond-laser-pulse-generation module (Figure 1a) and a fluidic chamber above the laser module, in which cells are lysed (Figure 1b). The microsecond laser pulse was generated by a 980-nm diode laser with a maximum power of 800 mW (Laserlands, 980MD-0.8W-BL, Hengelo, The Netherlands), and then modulated using a function generator (Agilent 33220A, Keysight Technologies, Santa Rosa, CA, USA) to control the on and off cycles via a transistor-transistor logic (TTL) pulse signal. The pulse frequency is set at 50 Hz, while the pulse width can also be controlled precisely for various cell lysis tests. A $10\times$ objective lens was applied to focus the laser onto the bottom of the fluidic chamber from beneath, as a 4.4 µm-diameter light spot, corresponding to a measured intensity of 508 kW/cm^2. Since the diode laser was mounted on an X-Y stage, it can be freely moved to locate the laser spot at any position on the bottom of the fluidic chamber bottom for targeting cells.

The fluidic chamber for cell lysis consists of a 1-mm-thick glass slide (top) and an optically-absorbent substrate (bottom). The fluidic chamber was filled with biocompatible solutions as the working media, in which the cells can be cultured and lysed. The optically-absorbent substrate is a 1-mm-thick glass slide, with a 200-nm-thick layer of indium tin oxide (ITO), topped with a 1-µm-thick layer of amorphous silicon (α-silicon). These absorbing materials help the bottom substrate absorb approximately 70% of the incident light from the laser [25], which is converted into heat that induces the vapor microbubbles in the fluidic chamber at the position of the laser spot on the substrate. The top and bottom of the chamber are separated by uniform-sized polystyrene beads (Polysciences, Inc., Warrington, FL, USA) with desired diameters, allowing discrete adjustment of the chamber height. Spacers were put on two opposite sides of the chamber, leaving the other two sides open for the fluid exchange.

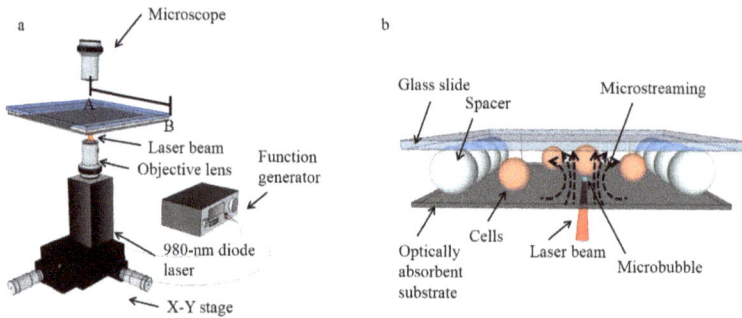

Figure 1. Setup for the single-cell lysis system. (**a**) A 980-nm diode laser was mounted on an X-Y stage, and modulated by a function generator to produce microsecond laser pulses. The laser light was focused by an objective lens onto the optically-absorbent bottom of the fluidic chamber, creating microbubbles that oscillated in size in the microfluidic chamber. (**b**) 3D structure of the microfluidic chamber filled with biocompatible solutions and consisting of an optically-absorbent substrate, a chamber ceiling made of a glass slide, and polystyrene beads acting as spacers. The cells can be cultured and lysed in the fluidic chamber.

2.2. Mechanism

The light from the focused laser spot on the optically absorbent substrate was transformed into heat, creating a microscale vapor bubble on the bottom of the fluidic chamber. The microbubble rapidly expands when the laser is on, and collapses when the laser is off. This process occurs repeatedly as the laser is pulsed. The size oscillation of the microbubble induced microstreaming around the bubble, corresponding to a strong shear stress. As shown in the Figure 1b, there is a rapid flow in the vertical direction caused by the microbubble oscillation [21,26]. Therefore, the targeted cell above the bubble experiences sufficient shear stress to rupture the cell membrane [17,27]. Another important factor for cell lysis is the direct contact of the cell membrane with the expanding microbubble [28,29]. The expanded bubble can be large enough (diameter of 7 to 14 μm) to contact the cell membrane positioned above the bubble, rupturing the membrane. If the induced microbubble is not large enough to touch the cell membrane, the lysis yield is dramatically decreased. The repeated expanding and collapsing cycles of the microbubble help lyse the whole cell membrane, while one cycle is sufficient to lyse the cell partially. The detailed cell lysis process was recorded with a high-speed camera at a frame rate of 200 fps (Figure 2). The whole cell lysis process lasted 400 ms, during which the membrane of the targeted cell was repeatedly ruptured by the bubble until the cell membrane was completely lysed.

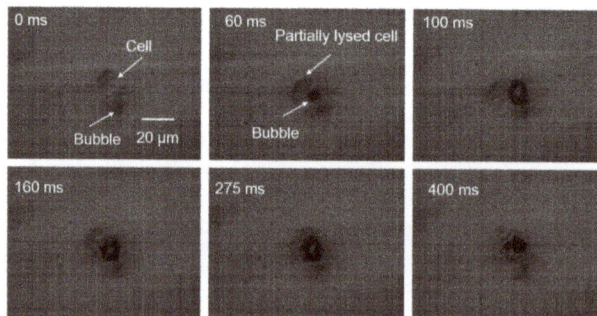

Figure 2. Cell-bubble interaction in one single-cell lysis test. Optical images were taken over a period of 400 ms, corresponding to the length of the cell lysis procedure, at a frame rate of 200 fps.

3. Materials and Methods

3.1. Cell Culture

NIH/3T3 (murine fibroblasts, ATCC, Manassas, VA, USA) were cultured in Dulbecco's Modified Eagle's Medium (DMEM, ATCC), containing 10% bovine serum (Gibco, Invitrogen, Carlsbad, CA, USA), penicillin (100 U/mL), and streptomycin (100 μg/mL). Cells were maintained at 37 °C in a humidified atmosphere of 5% CO_2 in air. The medium was replaced every 2–3 days. Immediately before cell lysis tests, 1 mL of 0.25% (w/v) Trypsin-0.53 mM EDTA solution was added to 25 cm^2 cell culture flask (Falcon, Corning Life Sciences, Corning, NY, USA) to detach the cell monolayer. Then cells were aspirated by gentle pipetting to obtain a cell suspension. The acquired cell suspension was added into microfluidic chamber for the single-cell lysis.

3.2. Cell Lysis

Prepared cell suspensions were added into the fluidic chamber, and cells dispersed on the bottom of the fluidic chamber. The laser can be moved freely using the X-Y stage to target a specific single cell. Once the position of laser and the targeted cell overlapped, the modulated laser pulses were triggered, creating the rapidly expanding cavitation microbubble to lyse the targeted cell.

Calcein AM (Invitrogen) is a green fluorescent dye that can penetrate the membrane of live cell, and emits a green fluorescence when it is hydrolyzed by live cells. If the membrane of a cell containing Calcein AM is ruptured, the cell interior will diffuse into the surrounding medium, and this process can be tracked by monitoring the green fluorescence of the Calcein AM dye. Therefore, prior to the experiment, cells were stained with Calcein AM to indicate cell lysis performance and the distribution of cell cytosol following lysis.

4. Characterization

The lysing efficiency can be affected by various experimental conditions, which were studied to maximize the single-cell lysis rate. The parameters that were varied include: laser pulse width, height of the microfluidic chamber, and the lysis duration for each cell. Thirty cells were tested for each parameter in the following experiments.

4.1. Laser Pulse Width

A longer laser pulse width can induce larger microbubbles, which have higher cell lysis yields. As shown in Figure 3a, the bubble size gradually grows from 7.4 μm ± 0.4 μm to 14.2 μm ± 1.0 μm when the laser pulse width increases from 300 μs to 1.2 ms. With the increasing bubble size, the cell lysis yield is also dramatically increased from 13.3% ± 13.3% to 97.0% ± 3.0% (Figure 3b). It can be seen that, for the 1.2-ms laser pulse width, induced microbubbles with the diameter of approximately 14 μm result in a high cell lysis yield. The tests for various laser pulse widths were all conducted in a 15-μm high microfluidic chamber, using a lysis duration of 400 ms. Thus, when the bubble size is close to the chamber height, there is a higher lysis yield, since the bubble has more chances to contact and rupture the cell membrane. The microstreaming around larger oscillating bubbles also induces strong shear stresses that push the targeted cell against the ceiling of the chamber top, followed by lysis. If the laser pulse is too short, the induced bubble is too small, and may not be able to reach the cell membrane. If the laser pulse width is too long, the size of the induced microbubble will be too large, making it difficult to lyse single cells precisely without damaging nearby cells.

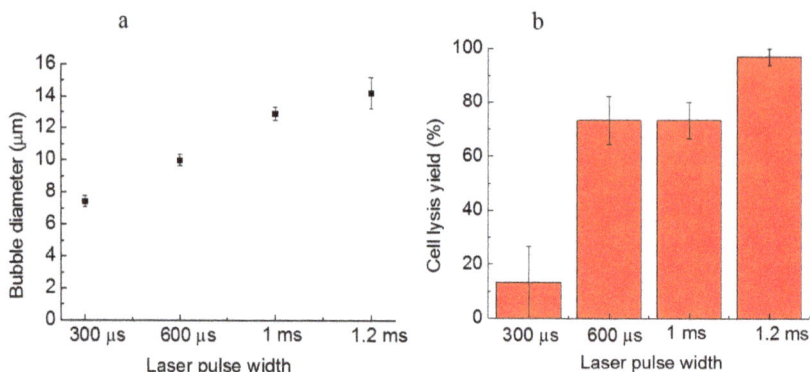

Figure 3. (**a**) The size of microbubbles induced with various laser pulse widths; (**b**) the cell lysis rate as a function of the laser pulse width. More than 30 cells were tested in three parallel experiments. Error bars show the standard error of the measurements.

4.2. Microfluidic Chamber Height

The cells can be vertically confined by the microfluidic chamber ceiling to keep them in place while the oscillating microbubble applied shear stress to the cell. Various chamber heights result in different cell lysis performances, depending on how tightly the chamber confines the targeted cells. As the distance between the bubble and the cell membrane increased, the shear stress decreased dramatically. Thus, a controlled chamber height can optimize the working distance between the microbubble and the targeted cell, increasing cell lysis yield.

Empirical observations showed that the minimum chamber height that allowed cell movement in the lateral direction in the fluidic chamber during lysis was 10 μm, which indicates that the cell diameter in the vertical direction is approximately 10 μm. The diameter of NIH3T3 cells was measured to be 16.6 μm ± 1.0 μm (averaged over 10 cells) from the top view when the cells rested on the substrate. However, due to gravitational force, cells in the chamber should have a height less than the diameter, which matches observations.

As described in the setup section, the chamber height can be controlled by the spacer thickness, using polystyrene beads of 10, 15, and 20 μm, respectively (Figure 4). The chamber height of 10 μm provides the highest cell lysis rate, and is similar to the lysis rate of 97.0% ± 3.0% for a 15-μm chamber height. When the chamber height was increased further to 20 μm, the cell lysis yield dropped to less than 20%, due to lack of vertical and lateral confinement of the cells, making it easy for the cells to escape the microstreaming surrounding the bubble. Cells move more freely in the lateral direction in the 15-μm-high fluidic chamber than in the 10-μm-high chamber, which facilitates cell transportation and patterning. Thus, the chamber height of 15 μm was chosen as the optimal fluidic chamber height. In these tests, the laser pulse width was 1.2 ms, and the lysis duration was 400 ms.

Heat produced from the laser absorption in the substrate has previously been characterized [21]. The laser focal point in this work was less than 1 μm in diameter, so even at the maximum laser power, the temperature 14.5 μm away from the center of the laser focal point is less than 32 °C. Since the optimal chamber height for cell lysis was 15 μm, the cells are at a safe temperature, even if the bubble is directly contacting the cell.

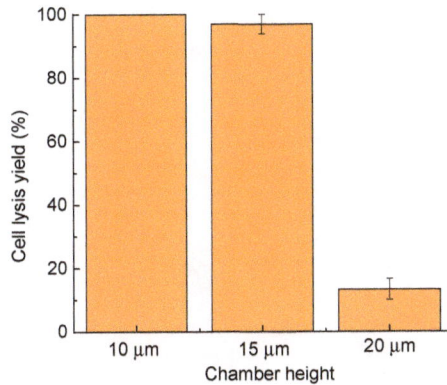

Figure 4. Cell lysis yield as a function of the chamber height. Thirty cells were tested in three parallel experiments for each chamber height. Error bars show the standard error of the measurements.

4.3. Cell Lysis Duration

A longer duration for the cell lysis process enhances the cell lysis yield. However, some studies of highly dynamic processes in cells, like the research on the activity of certain important cell signaling kinases using enzyme-specific peptide reporters [30–32], require high temporal resolution, which needs a shorter lysis duration. Therefore, an optimized cell lysis duration minimizes the lysis time per cell while maintaining a high cell lysis yield. As shown in Figure 5, the cell lysis yield can reach to 97.0% ± 3.0% as the cell lysis duration increases from 20 to 400 ms. The targeted cells exposed to laser-pulse-induced microbubbles for 200 ms had thoroughly ruptured membranes at a yield of 93.3% ± 3.3%. As the lysis duration increased to 400 ms, the cell lysis yield also increased, to 97.0% ± 3.0%. Therefore, 200-ms to 400-ms lysis durations result in high cell lysis yields.

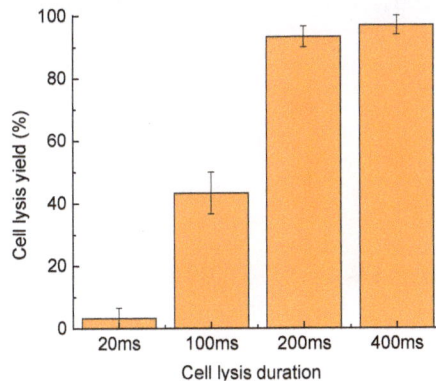

Figure 5. Cell lysis yield as a function of the cell lysis duration. Thirty cells were tested in three parallel experiments. Error bars show the standard error of the measurements.

5. Results and Discussion

5.1. High-Resolution Single-Cell Lysis

Localized single-cell lysis can be realized precisely using the optimized parameters for this system. To demonstrate high-resolution single-cell lysis, the cell interior was stained with Calcein AM prior to lysis. If the cell membrane was intact, the cell interior was confined, and showed an intense green fluorescence. Once the cell was lysed, the green fluorescence dissipated as the cell interior diffuses out into the solution. As shown in Figure 6, the targeted cell (Figure 6a) was lysed successfully, indicating by the ruptured cell membrane (Figure 6b), and subsequent decrease in the green fluorescence (Figure 6d). The cell adjacent to the targeted cell was not affected.

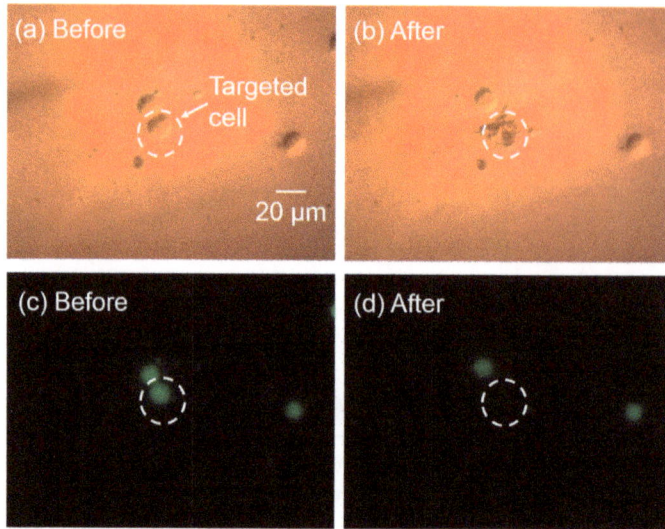

Figure 6. Targeted single-cell lysis result. (**a,b**) Differential interference contrast (DIC) images of cells before and after the targeted single-cell lysis. The targeted cell membrane was ruptured, while the nearby cell was kept intact. (**c,d**) The corresponding fluorescent images before and after cell lysis. The green fluorescence of the targeted cell dissipated, while the nearby cells maintained their intense green fluorescence, indicating the successful lysis of targeted single cell with no effect on the neighboring cells.

5.2. Dilution of Cell Content

To determine the cell content distribution change after the cell lysis, the cell contents were stained with Calcein AM prior to the tests. The cell was placed in the center of an area encompassing 100 μm × 100 μm. The fluorescent images before cell lysis and 5 s after cell lysis were examined to measure the change of the amount of fluorescent dye, represent the spatial distribution of the cell content. The integrated fluorescence of the examined area was measured with the software ImageJ after the background fluorescence was subtracted. The integrated fluorescence after the cell lysis was compared with the one before cell lysis to quantify the dilution of the cellular contents (Figure 7). After cell lysis, the Calcein AM diffuses, so the integrated fluorescent intensity decreased gradually. However, there was still 85.3% of the fluorescent dye in the examined area surrounding the lysed cell 2 s post-lysis, and 56.5% of the dye left after 5 s, indicating that with this single-cell lysis technique, there is enough time left to collect or analyze the lysate before it disperses widely. Many useful components of the lysate are larger than Calcein AM, like large proteins, genetic materials, or organelles, making these even easier to collect after lysis.

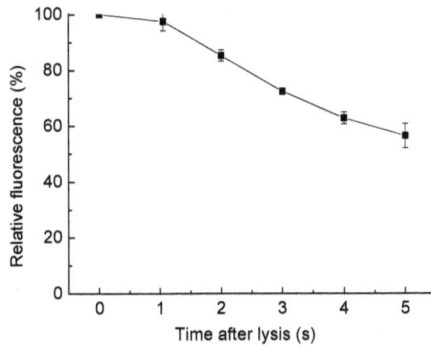

Figure 7. Dilution of cell content as the time passed after cell lysis. Relative fluorescence is measured as the integrated fluorescent density compared with it before the cell lysis, in a 100 μm × 100 μm area surrounding the targeted cell, with the background subtracted. Since the cell interior was stained with Calcein AM, the fluorescence change represents the dilution of cell content. The data is from three parallel tests, and error bars show the standard error of the measurements.

5.3. Lysis of a Subcellular Region

By controlling the laser pulse width and cell lysis duration, the cell membrane can be ruptured only in a subcellular region. This high-resolution cell lysis can help biological studies, since many biochemical reactions happen in spatially-discrete regions of cells, like the signaling proteins present at the neuronal synapses and growth cone [33]. Thus, lysis in a localized region of the cell can help realize subcellular analysis [33]. As shown in Figure 8a–g, when subcellular lysis occurs, the cell membrane has a localized opening, causing the cell interior to gradually leak out. Figure 8a,b shows the cell morphology change after the subcellular lysis. The membrane was partially ruptured, and a hole was generated. The fluorescent images in Figure 8d–g show the Calcein-AM-stained cell interior gradually leaked out in 12 s after the cell lysis. Figure 8c shows the relative fluorescence of the target lysed cell, measured over a 100 μm by 100 μm area bounding the cell. This result shows the cell contents are confined in the vicinity of the target cell within 5 s, similar to the results for complete lysis.

Cell lysis in a subcellular region can help with analyzing a specific area of a cell, but it will not release all of the cellular contents completely and rapidly. Though longer laser pulse widths or cell lysis durations can induce lysis over a larger area of the cell membrane, until the whole cell membrane is ruptured, there is always a compromise between lysis resolution and lysis efficiency. Figure 9a,b shows the effect of the laser pulse width and the cell lysis duration on lysis. The bottom portion of each bar represents subcellular lysis, while the upper portion of each bar represents the lysis of the entire cell membrane. The whole bar equals the total cell lysis yield. If the cell was significantly disrupted (more than 50% of the cell shape is disrupted), it was categorized as whole cell lysis, while if more than 50% of the cell shape was maintained, with only a hole appearing in the membrane, it was categorized as subcellular cell lysis. The laser duration was kept at 400 ms to characterize the various pulse width effects, as shown in Figure 9a. For a pulse width of 300 μs, the subcellular lysis approaches 96.7%. When the pulse width was increased to 600 μs, the entire membrane lysis dominates subcellular lysis. Therefore, the pulse width of 300 μs was chosen for subcellular lysis. Another parameter, lysis duration, was varied from 20 to 400 ms (Figure 9b), while keeping the laser pulse width at 1.2 ms. Since the pulse frequency is 50 Hz, only a single laser pulse occurs in the minimum duration of 20 ms, but this is sufficient to induce subcellular lysis. When the lysis duration was lengthened, cells were more likely to be completely lysed. If only the total cell lysis yield is considered, without bias to subcellular or entire cell membrane lysis, the 300-μs laser pulse width and 20-ms lysis duration have the highest cell lysis yield of 100%.

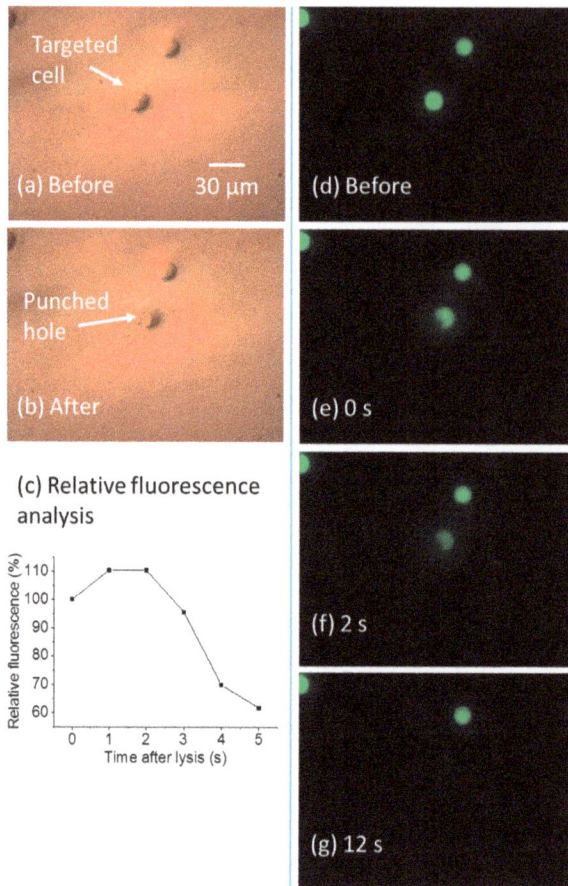

Figure 8. Lysis in a subcellular region. (**a,b**) DIC images before and after the subcellular lysis; (**c**) relative fluorescence of the target cell after lysis; (**d–g**) fluorescent images before the lysis, and up to 12 s after the lysis.

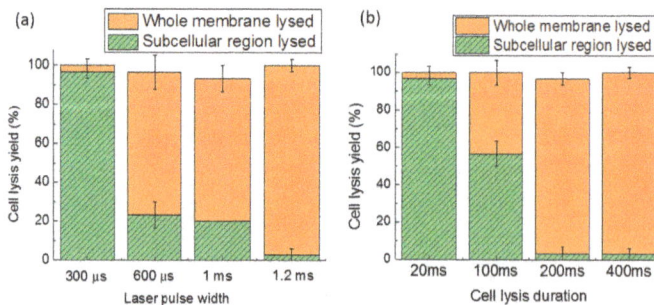

Figure 9. Cell lysis yield as a function of the laser pulse width (**a**); and cell lysis duration (**b**). Thirty cells were tested in three parallel experiments. Error bars show the standard error of the measurements.

5.4. Integration of Cell Manipulation and Cell Lysis

Integration of a cell lysis system to other devices is an important requirement for cell analysis [2,10]. In this single-cell lysis system, it is also possible to perform cell manipulation, cell poration, on-site cell culture, and cell content analysis [21,34]. Cell manipulation prior to cell lysis is useful to studies related to cell–cell interactions, like the influence of stem cells on surrounding fibroblasts [35]. Prior to cell lysis, selected cells can be patterned and co-cultured on-chip for creating cell–cell interactions. A targeted cell in the cell pattern can be lysed precisely, and the lysate can be collected by capillary electrophoresis or analyzed on site via the fluorescent reactions in the medium for the single-cell analysis [18,36].

The integration of the cell patterning and cell lysis in the same microfluidic device has been demonstrated with this single-cell lysis system. As shown in Figure 10, three cells were randomly located. The green fluorescence from Calcein AM in the cell interior indicates intact cell membranes and cell activity. Laser pulse-induced microbubbles were used to manipulate the cells and pattern them into a line via the "pulling mode" of the microbubbles [21] in less than 15 s. This pulling mode creates forces that attracts the cells towards the laser focal point [21], and is achieved by setting the laser pulse width to 7 μs. After cell patterning, the green fluorescence of the cells was maintained, indicating the cell membrane was not affected during the patterning process. When the cell patterning was finished, the left cell was chosen to be lysed. The result in Figure 10e,f shows that the targeted cell was successfully lysed without affecting neighboring cells. This process of cell patterning and cell lysis was fulfilled precisely in the same chip and same place, providing a practical method for biological studies.

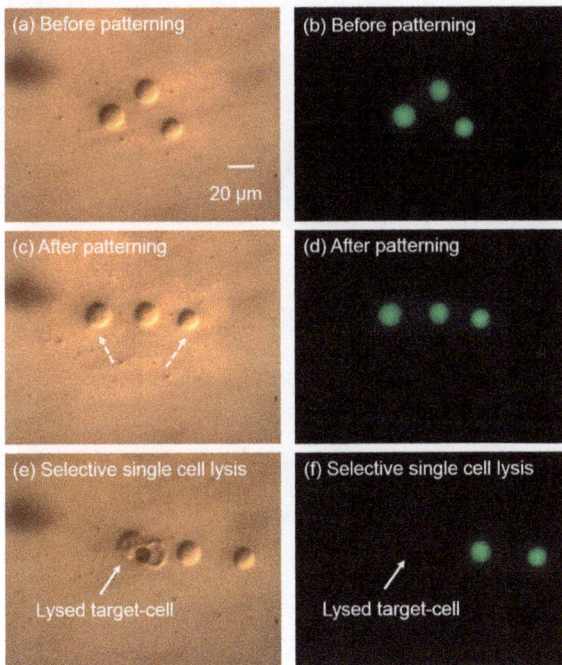

Figure 10. Cell patterning and single cell lysis on the same chip. The left column of images shows the brightfield images, while the right column shows the corresponding fluorescent images. (**a,b**) Cells randomly positioned before patterning; (**c,d**) cells were patterned into a line with the laser-induced microbubbles. The dashed arrows show the cell trajectories during the cell patterning. (**e,f**) The targeted cell was lysed selectively, without affecting neighboring cells.

5.5. Adherent Single-Cell Lysis

This single-cell lysis system works both for suspension cells and adherent cells. The cell lysis yield for adherent cells can reach to 100.0% ± 0.0% using the previous optimized parameters, measured in 30 tests. Figure 11 shows a demonstration: after the precise single-cell lysis, the cell membrane was ruptured, and the fluorescent-dye from the cell interior leaked into the medium, while the neighboring cells were not affected.

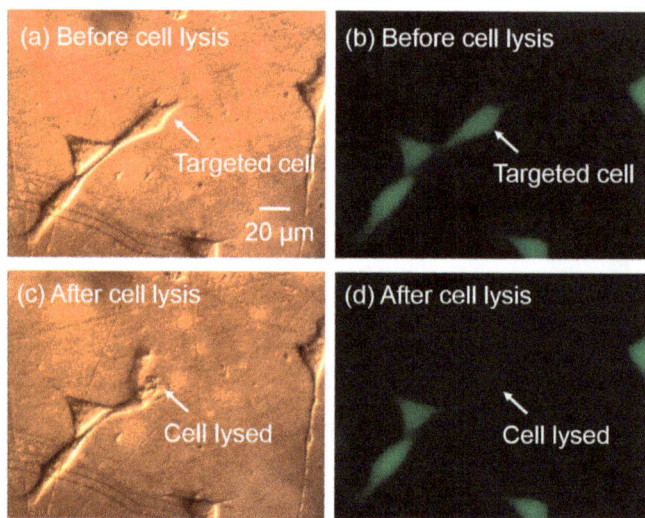

Figure 11. Adherent single-cell lysis. Images in left column are brightfield images, while images in the right column are the corresponding fluorescent images. (**a**,**b**) Before cell lysis, the cell membrane was intact and showing green fluorescence; (**c**,**d**) after cell lysis, the targeted cell was successfully lysed without affecting the neighboring cells.

6. Conclusions

A localized single-cell lysis system using microsecond laser pulses was demonstrated. The microsecond-laser-induced size-oscillating microbubbles apply strong shear stress to the targeted cell right above the bubble. In addition, the rapidly expanding and collapsing bubbles repeatedly directly contact the cell membrane and rupture the membrane. These two factors cause the cell membrane rupture, and complete lysis at a yield of up to 97.0% ± 3.0%. Various laser pulse widths, chamber heights, and cell lysis durations were characterized to obtain higher cell lysis efficiency. Localized single cells can be lysed completely and precisely without affecting the neighboring cells. The lysis can achieve a subcellular spatial resolution. The lysis process takes 300 µs per cell for subcellular membrane lysis and 200 to 400 ms per cell for lysis of the entire cell. After cell lysis, the dilution rate of the of cell content enables subsequent cell lysate collection or on-site analysis, since small molecules in the cell content, such as the fluorescent dye Calcein AM, still retain 85.3% of the original concentration within the surrounding 100 µm × 100 µm area up to 2 s after lysis. Cell manipulation functions can also be integrated into this system, so cells can be freely patterned and co-cultured, and then be selectively lysed for cell analysis. This single-cell lysis system works with suspension cells and adherent cells.

This single-cell lysis system can be adopted for many biomedical studies, since the experimental setup is easy to acquire and use. There is no need to fabricate complex microfluidic structures on the substrate, and all of the cells in the microfluidic chamber can be selectively lysed with high resolution. Cell lysis and manipulation can occur in a chamber with no microstructures, providing a platform for tissue engineering applications. For example, specific cells can be patterned for tissue growth, followed by the lysis of selected single target cells. This can be used to study cell–cell interaction within the tissue. In the future, the parallel and automated control of microbubbles will be realized, enabling the lysis of multiple target cells at the same time, thus further increasing the cell lysis throughput. This can be realized by a laser scanning system that can project a single laser onto multiple areas of the substrate within one period, as the laser pulse width of 300 μs to 1.2 ms is far smaller than the 20-ms pulse period that was used [25].

Acknowledgments: This project was supported in part by Grant Number 1R01EB016458-01 from the National Institute of Biomedical Imaging and Bioengineering of the National Institutes of Health (NIH) as part of the NSF/NASA/NIH/USDA National Robotics Initiative. These contents are solely the responsibility of the authors and do not necessarily represent the official views of the NIH.

Author Contributions: Qihui Fan, Wenqi Hu, and Aaron T. Ohta conceived of the system and its operation, and designed the experiments; Qihui Fan and Wenqi Hu performed the experiments; Qihui Fan and Aaron T. Ohta analyzed the data; Qihui Fan wrote the manuscript; and Qihui Fan, Wenqi Hu, and Aaron T. Ohta edited the manuscript.

Conflicts of Interest: The authors declare no conflict of interest.

References

1. Di Carlo, D.; Lee, L.P. Dynamic single-cell analysis for quantitative biology. *Anal. Chem.* **2006**, *78*, 7918–7925. [CrossRef] [PubMed]
2. Brown, R.B.; Audet, J. Current techniques for single-cell lysis. *J. R. Soc. Interface R. Soc.* **2008**, *5* (Suppl. S2), S131–S138. [CrossRef] [PubMed]
3. Jen, C.P.; Hsiao, J.H.; Maslov, N.A. Single-cell chemical lysis on microfluidic chips with arrays of microwells. *Sensors* **2012**, *12*, 347–358. [CrossRef] [PubMed]
4. Han, F.; Wang, Y.; Sims, C.E.; Bachman, M.; Chang, R.; Li, G.P.; Allbritton, N.L. Fast electrical lysis of cells for capillary electrophoresis. *Anal. Chem.* **2003**, *75*, 3688–3696. [CrossRef] [PubMed]
5. Hung, M.-S.; Chang, Y.-T. Single cell lysis and DNA extending using electroporation microfluidic device. *Biochip J.* **2012**, *6*, 84–90. [CrossRef]
6. Jen, C.P.; Amstislavskaya, T.G.; Liu, Y.H.; Hsiao, J.H.; Chen, Y.H. Single-cell electric lysis on an electroosmotic-driven microfluidic chip with arrays of microwells. *Sensors* **2012**, *12*, 6967–6977. [CrossRef] [PubMed]
7. Jokilaakso, N.; Salm, E.; Chen, A.; Millet, L.; Guevara, C.D.; Dorvel, B.; Reddy, B.; Karlstrom, A.E.; Chen, Y.; Ji, H.; et al. Ultra-localized single cell electroporation using silicon nanowires. *Lab Chip* **2013**, *13*, 336–339. [CrossRef] [PubMed]
8. Lin, Y.-H.; Lee, G.-B. An integrated cell counting and continuous cell lysis device using an optically induced electric field. *Sens. Actuators B Chem.* **2010**, *145*, 854–860. [CrossRef]
9. Kremer, C.; Witte, C.; Neale, S.L.; Reboud, J.; Barrett, M.P.; Cooper, J.M. Shape-dependent optoelectronic cell lysis. *Angew. Chem. Int. Ed. Engl.* **2014**, *53*, 842–846. [CrossRef] [PubMed]
10. Nan, L.; Jiang, Z.; Wei, X. Emerging microfluidic devices for cell lysis: A review. *Lab Chip* **2014**, *14*, 1060–1073. [CrossRef] [PubMed]
11. Zhang, H.; Jin, W. Single-cell analysis by intracellular immuno-reaction and capillary electrophoresis with laser-induced fluorescence detection. *J. Chromatogr. A* **2006**, *1104*, 346–351. [CrossRef] [PubMed]
12. Zhu, K.; Jin, H.; Ma, Y.; Ren, Z.; Xiao, C.; He, Z.; Zhang, F.; Zhu, Q.; Wang, B. A continuous thermal lysis procedure for the large-scale preparation of plasmid DNA. *J. Biotechnol.* **2005**, *118*, 257–264. [CrossRef] [PubMed]
13. Carlo, D.D.; Jeong, K.-H.; Lee, L.P. Reagentless mechanical cell lysis by nanoscale barbs in microchannels for sample preparation. *Lab Chip* **2003**, *3*, 287–291. [CrossRef] [PubMed]
14. Kim, J.; Hong, J.W.; Kim, D.P.; Shin, J.H.; Park, I. Nanowire-integrated microfluidic devices for facile and reagent-free mechanical cell lysis. *Lab Chip* **2012**, *12*, 2914–2921. [CrossRef] [PubMed]
15. Rau, K.R.; Guerra, A.; Vogel, A.; Venugopalan, V. Investigation of laser-induced cell lysis using time-resolved imaging. *Appl. Phys. Lett.* **2004**, *84*, 2940–2942. [CrossRef]
16. Rau, K.R.; Quinto-Su, P.A.; Hellman, A.N.; Venugopalan, V. Pulsed laser microbeam-induced cell lysis: Time-resolved imaging and analysis of hydrodynamic effects. *Biophys. J.* **2006**, *91*, 317–329. [CrossRef] [PubMed]

17. Hellman, A.N.; Rau, K.R.; Yoon, H.H.; Venugopalan, V. Biophysical response to pulsed laser microbeam-induced cell lysis and molecular delivery. *J. Biophotonics* **2008**, *1*, 24–35. [CrossRef] [PubMed]
18. Lai, H.-H.; Quinto-Su, P.A.; Sims, C.E.; Bachman, M.; Li, G.P.; Venugopalan, V.; Allbritton, N.L. Characterization and use of laser-based lysis for cell analysis on-chip. *J. R. Soc. Interface* **2008**, *5*, S113–S121. [CrossRef] [PubMed]
19. Quinto-Su, P.A.; Lai, H.-H.; Yoon, H.H.; Sims, C.E.; Allbritton, N.L.; Venugopalan, V. Examination of laser microbeam cell lysis in a PDMS microfluidic channel using time-resolved imaging. *Lab Chip* **2008**, *8*, 408–414. [CrossRef] [PubMed]
20. Li, Z.G.; Liu, A.Q.; Klaseboer, E.; Zhang, J.B.; Ohl, C.D. Single cell membrane poration by bubble-induced microjets in a microfluidic chip. *Lab Chip* **2013**, *13*, 1144–1150. [CrossRef] [PubMed]
21. Hu, W.; Fan, Q.; Ohta, A.T. An opto-thermocapillary cell micromanipulator. *Lab Chip* **2013**, *13*, 2285–2291. [CrossRef] [PubMed]
22. Zhao, C.; Xie, Y.; Mao, Z.; Zhao, Y.; Rufo, J.; Yang, S.; Guo, F.; Mai, J.D.; Huang, T.J. Theory and experiment on particle trapping and manipulation via optothermally generated bubbles. *Lab Chip* **2014**, *14*, 384–391. [CrossRef] [PubMed]
23. Thery, M. Micropatterning as a tool to decipher cell morphogenesis and functions. *J. Cell Sci.* **2010**, *123*, 4201–4213. [CrossRef] [PubMed]
24. Gautrot, J.E.; Trappmann, B.; Oceguera-Yanez, F.; Connelly, J.; He, X.; Watt, F.M.; Huck, W.T.S. Exploiting the superior protein resistance of polymer brushes to control single cell adhesion and polarisation at the micron scale. *Biomaterials* **2010**, *31*, 5030–5041. [CrossRef] [PubMed]
25. Hu, W.; Fan, Q.; Ohta, A. Interactive actuation of multiple opto-thermocapillary flow-addressed bubble microrobots. *Robot. Biomim.* **2014**, *1*, 14. [CrossRef] [PubMed]
26. Wu, J.; Nyborg, W.L. Ultrasound, cavitation bubbles and their interaction with cells. *Adv. Drug Deliv. Rev.* **2008**, *60*, 1103–1116. [CrossRef] [PubMed]
27. Marmottant, P.; Hilgenfeldt, S. Controlled vesicle deformation and lysis by single oscillating bubbles. *Nature* **2003**, *423*, 153–156. [CrossRef] [PubMed]
28. Handa, A.; Emery, A.N.; Spier, R.E. On the evaluation of gas-liquid interfacial effects on hybridoma viability in bubble column bioreactors. *Dev. Biol. Stand.* **1987**, *66*, 241–253. [PubMed]
29. Michaels, J.D.; Nowak, J.E.; Mallik, A.K.; Koczo, K.; Wasan, D.T.; Papoutsakis, E.T. Analysis of cell-to-bubble attachment in sparged bioreactors in the presence of cell-protecting additives. *Biotechnol. Bioeng.* **1995**, *47*, 407–419. [CrossRef] [PubMed]
30. Meredith, G.D.; Sims, C.E.; Soughayer, J.S.; Allbritton, N.L. Measurement of kinase activation in single mammalian cells. *Nat. Biotechnol.* **2000**, *18*, 309–312. [PubMed]
31. Li, H.; Wu, H.Y.; Wang, Y.; Sims, C.E.; Allbritton, N.L. Improved capillary electrophoresis conditions for the separation of kinase substrates by the laser micropipet system. *J. Chromatogr. B Biomed. Sci. Appl.* **2001**, *757*, 79–88. [CrossRef]
32. Li, H.; Sims, C.E.; Kaluzova, M.; Stanbridge, E.J.; Allbritton, N.L. A quantitative single-cell assay for protein kinase B reveals important insights into the biochemical behavior of an intracellular substrate peptide. *Biochemistry* **2004**, *43*, 1599–1608. [CrossRef] [PubMed]
33. Li, H.; Sims, C.E.; Wu, H.Y.; Allbritton, N.L. Spatial control of cellular measurements with the laser micropipet. *Anal. Chem.* **2001**, *73*, 4625–4631. [CrossRef] [PubMed]
34. Fan, Q.; Hu, W.; Ohta, A.T. Efficient single-cell poration by microsecond laser pulses. *Lab Chip* **2015**, *15*, 581–588. [CrossRef] [PubMed]
35. Hong, S.; Pan, Q.; Lee, L.P. Single-cell level co-culture platform for intercellular communication. *Integr. Biol.* **2012**, *4*, 374–380. [CrossRef] [PubMed]
36. Mellors, J.S.; Jorabchi, K.; Smith, L.M.; Ramsey, J.M. Integrated microfluidic device for automated single cell analysis using electrophoretic separation and electrospray ionization mass spectrometry. *Anal. Chem.* **2010**, *82*, 967–973. [CrossRef] [PubMed]

micromachines

MDPI

Article

A Micromanipulator and Transporter Based on Vibrating Bubbles in an Open Chip Environment

Liguo Dai [1,2], Niandong Jiao [1,*], Xiaodong Wang [1,2] and Lianqing Liu [1,*]

[1]   State Key Laboratory of Robotics, Shenyang Institute of Automation, Chinese Academy of Sciences, Shenyang 10016, China; dailiguo@sia.cn (L.D.); wangxiaodong@sia.cn (X.W.)
[2]   University of Chinese Academy of Sciences, Beijing 100049, China
*   Correspondence: ndjiao@sia.cn (N.J.); lqliu@sia.cn (L.L.);
    Tel.: +86-24-2397-0540 (N.J.); +86-24-2397-0181 (L.L.)

Academic Editors: Aaron T. Ohta, Wenqi Hu and Nam-Trung Nguyen
Received: 13 January 2017; Accepted: 7 April 2017; Published: 18 April 2017

Abstract: A novel micromanipulation technique of multi-objectives based on vibrating bubbles in an open chip environment is described in this paper. Bubbles were created in an aqueous medium by the thermal energy converted from a laser. When the piezoelectric stack fixed under the chip vibrated the bubbles, micro-objects (microparticles, cells, etc.) rapidly moved towards the bubbles. Results from numerical simulation demonstrate that convective flow around the bubbles can provide forces to capture objects. Since bubbles can be generated at arbitrary destinations in the open chip environment, they can act as both micromanipulators and transporters. As a result, micro- and bio-objects could be collected and transported effectively as masses in the open chip environment. This makes it possible for scientific instruments, such as atomic force microscopy (AFM) and scanning ion conductive microscopy (SICM), to operate the micro-objects directly in an open chip environment.

Keywords: vibrating bubble; micromanipulation; cell trapping; open chip environment

1. Introduction

The manipulation of individual micro-objects, which usually includes capture, transport, rotation, and isolation, is becoming a critical technology for micro-assembly and biomedical applications. At the micro-scale level, particles and cells can be moved and patterned by traditional methods, such as optical traps [1–4], dielectrophoresis forces [5–7], acoustic waves [8–11], and magnetic fields [12–15]. Moreover, many new techniques and tools have drawn more attention because of their great potential in lab on a chip applications, one of which being microbubbles. Many researchers have focused on the novel and special function of bubbles in microfluidic systems, such as pumps [16,17], actuators [18,19], mixers [20,21], and valves [22,23], as well as in other devices [24–27].

Bubbles have become a versatile tool for various lab on a chip and micromanipulation applications, and many innovative devices have been explored in recent years [28]. Microbubbles can be used as focusing agents when the acoustic streaming flow exerts a large enough shear force on vesicles in the flow, so it is possible to be developed for cell manipulation, cell-wall permeation and microfluidic devices [29]. Further, oscillating bubbles are capable of size- and density-based selective trapping of particles [30]. With other on-chip manipulation methods, both millimeter- and micro-sized objects could be captured, carried, and released by oscillating mobile bubbles [31]. Particle trapping and manipulation can also be completed by optothermally-generated bubbles, rather than oscillating bubbles [32]. Bubbles used for manipulation are generally controlled by the electrowetting-on-dielectric (EWOD) [31,33–35], or optical methods [32,36–39]. In an aqueous medium, bubbles transported by AC-electrowetting, and oscillated by piezo-actuator, are capable of capturing, carrying, and releasing objects [31,33]. Using two arrays of EWOD electrodes, twin bubbles driven and transported simultaneously can attract and capture beads and fish eggs [34,35]. Because bubbles are actuated by alternating the EWOD chip, the trajectory of bubbles and micro-objects is limited by design of the chip. Thus, another mechanism of generation and control of bubbles is proposed. In fluidic oil chambers, micro bubbles act as microrobots for manipulation and assembly, which are controlled by optically-induced heating and thermocapillary effects [36]. Micro-object trapping and manipulation can be completed by an optothermal bubble, due to the convective flow, surface tension, and pressure forces [32]. When a bubble is generated at the top of a gold film, a convective flow was formed around the bubble, so that a particle can be moved towards the bubble by the convective flow induced by a temperature gradient. After the particle approaches the bubble, it is trapped on the bubble's surface, because of the balance of pressure force and surface tension force along the radial direction of the bubble. Micro-objects can be carried along with the optically-controlled bubble to any desired location, while the closed fluidic chamber has limited the cooperation with other scientific instruments and technology. In addition, manipulation of the micro-particles and cells can also be completed with a non-contact method, where disk-shaped hydrogel microrobots actuated by laser-induced cavitation bubbles are used to draw patterns of cells and microgels [37]. Since the optothermally-induced fluid flow can trap and transport bio-objects, a micromanipulation platform based on bubbles is capable of manipulation and patterning [38]. However, since each microrobot or manipulator can only pattern and manipulate a single cell at a time, efficiency remains to be improved.

Here we present a novel manipulation method based on vibrating bubbles in an open chip environment, by which micro-objects (microparticles, cells, etc.) could be collected and transported efficiently as mass in the open chip environment. Both optothermally-generated bubbles [40] and oscillating bubbles [41] are popularly used in lab on a chip community, and have been well-studied for micromanipulator applications. However, using both techniques in one device is proposed for the first time in this paper. Compared with the methods of micro-objects manipulation mentioned above, this new technique can collect and move particles, as well as cells, in an open micro chamber without top glass. This technology is expected to function as a transporter for particle manipulation and transportation, collaboratively used in unclosed operating environments of other scientific equipment (atomic force microscopy (AFM) and scanning ion conductive microscopy (SICM), etc.).

2. Materials and Methods

2.1. Materials

In our experiment, the chip consisted of polydimethylsiloxane (PDMS, SYLGARD184, Dow Corning Holding Co., Ltd., Midland, MI, USA), glass substrate and gold layer, and the pre-polymer of PDMS was a mixture of base and curing agent. Purified deionized water was filled in the reservoir of chip when the manipulation objects were microballs. The balls with diameter of 50 to 100 µm were made of barium titanate glass (BTG). In the experiment of manipulating cells, Human Embryonic Kidney (HEK) 293 cells and pandorina morum cells acted as micro-objects. Experimental strains of HEK 293 cells were obtained from China Center for Type Culture Collection (Wuhan, China), and the strains of *pandorina morum* were provided by the Freshwater Algae Culture Collection at the Institute of Hydrobiology (Wuhan, China). The fluid in the reservoir was replaced by culture medium (Eagle's Minimum Essential Medium and Tris-acetate-phosphate medium respectively) in the experiments of manipulating cells with oscillating bubbles. The diameter of HEK 293 cells was about 20 µm, while the diameter of morum cells was 30 µm.

2.2. Methods

The new method can manipulate and trap multi-objects and cells at an arbitrary destination from relatively long distances away on the chip, and then transport them to a new location by another optothermally-generated bubble. As shown in Figure 1, to manipulate the micro-objects, a bubble is created on a chip coated with a gold layer. The diameter of the bubble is related to the intensity and irradiation time of the laser, for the bubble is produced by optically-induced heating. When the bubble is vibrated by a piezoelectric stack, objects were attracted to the bubble by convective flow. Using a high-voltage signal, the working distance of the micro bubbles may reach the millimeter scale. Theoretical analysis and simulations were conducted in our studies that reveal that micro-objects are driven towards the bubble vibrated by the piezo-actuator by heat-induced convective flow. If we turned on the laser, the bubble would increase in size continuously and explode, causing the micro-objects collected previously to disperse. When the frequency of the wave applied to the piezo-stack was transformed to the bubble's resonance frequency, the bubble could also be damaged. By changing the position of the next bubble after the previous one has burst, the dispersed micro-objects could be re-collected and moved to the new destination. Further, the moving distance of the particles could be as long as the channel in the chip. Simultaneous manipulation and transportation of multitarget objectors could be completed in an unclosed chip.

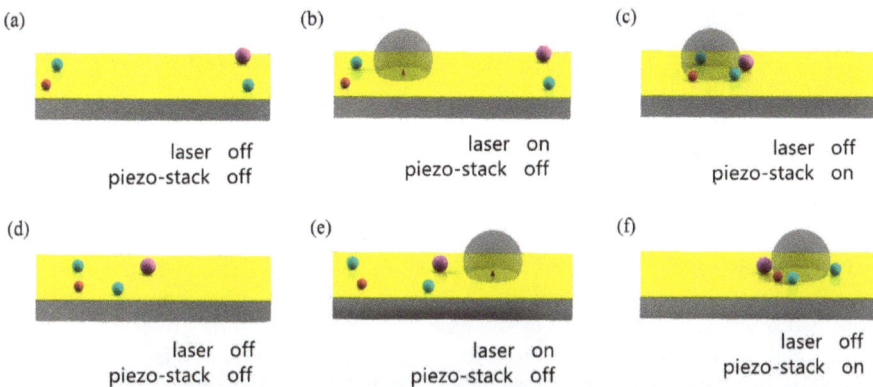

Figure 1. Collecting and transporting of micro-objects by oscillating vibrating bubbles: (**a**) micro-objects distributed on the chip; (**b**) a bubble generates on the chip; (**c**) piezoelectric stack is turned on and the particles are collected by the oscillated bubble; (**d**) the bubble bursts and the micro-objects disperse; (**e**) another bubble appears on the chip; (**f**) the new bubble collects these objects again.

2.3. Experiment Setup

The experiment system is shown in Figure 2, where a semiconductor laser (405 nm wavelength, 0–400 mW power), and a lens (25X, NA = 0.40), were used to provide sufficient power for the generation of a bubble. The laser and lens were fixed to a manual stage so that the position of the bubble generated was controllable and variable. A piezoelectric stack (PK2FMP2, Thorlabs Inc., Newton, NJ, USA), driven by an arbitrary waveform generator (ArbStudio 1102, Teledyne LeCroy Inc., Chestnut Ridge, NY, USA), together with an amplifier (33502A, Keysight Technologies Inc., Palo Alto, CA, USA), vibrated the micro bubble on the chip. The drive voltage of the piezo-actuator range was 0–75 V, and the displacement at 75 V was 11.2 μm. The chip was made up of a 1.2-mm-thick slide glass, and a small PDMS reservoir. A 50 nm thin-film layer gold layer was sputtered on the glass to absorb and transfer the energy of the laser beam. Other devices in this system included an optical microscope (1-60191D, Navitar Inc., Rochester, NY, USA), a camera (FL2G-13S2, Point Grey Research Inc., Richmond, BC, Canada), a computer, and a long pass filter (FELH0450, Throlabs, Newton, NJ, USA) with a 450 nm cut-on wavelength, which can reject the laser light into the microscope.

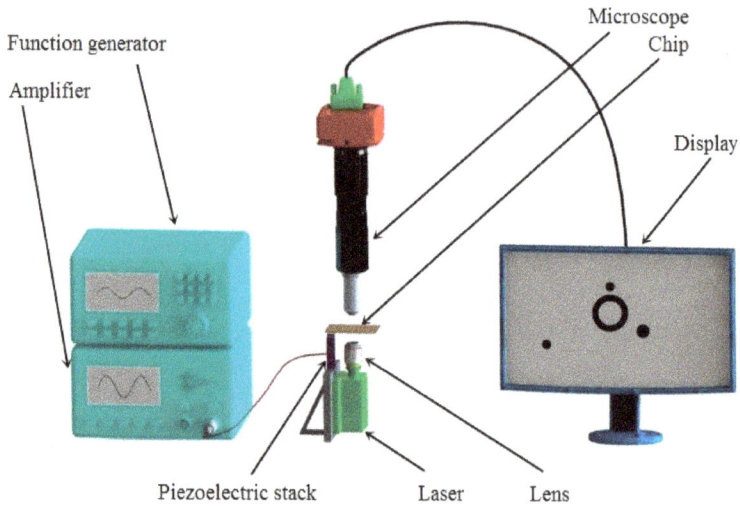

Figure 2. Schematic of the system setup.

2.4. Fabrication of Chip

The microfluidic chip, consisting of a glass substrate, fluid reservoir, and gold layer, has a simple design and can be fabricated rapidly. The reservoir was manufactured with PDMS, an elastomeric material [42,43]. Because of its physical and chemical properties, such as transparency, insulation, and nontoxicity, PDMS has become one of the most actively developed polymers for microfluidics. In contrast to general microfluidic chips, the chips used in these experiments were unclosed. The manufacturing process can be divided into five steps, as illustrated in Figure 3a–e. First, an acrylic mold is designed in a computer-aided design program and produced with machine tools. A pre-polymer of PDMS in the liquid state is then injected into the mold and cures gradually at a temperature of 75 °C. In our experiments, the PDMS included two ingredients—a base and a curing agent. An elastomeric and cross-linked solid was generated when the vinyl groups of the base reacted with the silicon hydride groups of the curing agent. These two kinds of solution were mixed in a mass ratio of 10:1 to produce a replica. Approximately four hours later, the liquid pre-polymer solidified and conformed to the shape of the master. The solidified PDMS cast was then peeled away from the die. Following this, the PDMS structure was oxidized for five minutes and sealed tightly and irreversibly to the slide glass. Silanol groups formed on the surface of the PDMS by the oxidation of methyl groups so that it could seal to a range of materials other than itself, including glass, silicon, and polyethylene. Since the gold layer prevented the linkage of PDMS and glass, the last procedure is the sputtering of gold on the chip. The diameter and depth of the reservoir on the fabricated chip, shown in Figure 3f, is 3 mm and 2 mm, respectively.

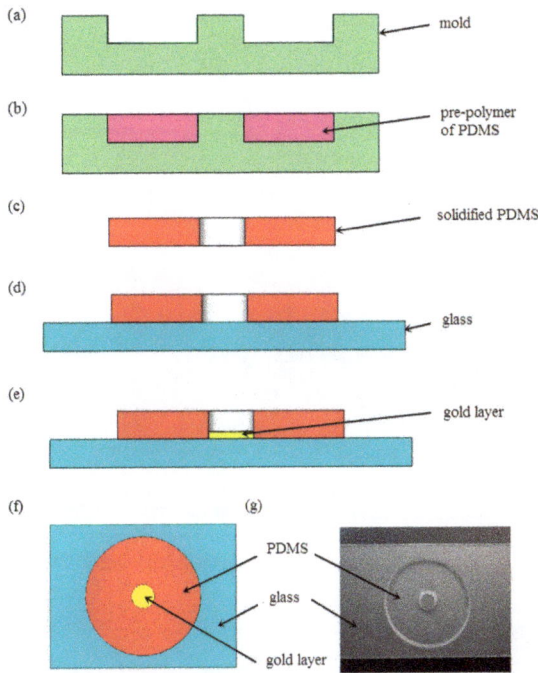

Figure 3. (**a–e**) Scheme of the fabrication process of the chip: (**a**) fabricating a mold; (**b**) pouring liquid pre-polymer into the pattern die and heating; (**c**) removing the cured polydimethylsiloxane (PDMS) copy; (**d**) combining the PDMS with a slide glass; (**e**) gold layer is sputtered on the glass; (**f**) schematic top view of the chip; (**g**) actual picture of the chip without gold layer.

3. Results and Discussion

3.1. Simulation

To provide theoretical guidance for the object trapping and manipulation process, computational fluid dynamics simulations were conducted using ANSYS Fluent software (Version 14.0, Pittsburgh, PA, USA). The goal of the simulations was to reveal the convective flow pattern around the oscillated bubble when the piezo-actuator was on. Micro-objects and cells could be moved by the force of the fluidic streaming, which is studied by experiments and numerical analysis. According to Navier-Stokes Equations [29,44], the motion of a viscous incompressible fluid can be described as:

$$\frac{\partial \rho}{\partial t} + \nabla(\rho v) = 0 \tag{1}$$

where ρ is the mass density, t is time, and v is the fluid density. In the two-equation turbulence model [45–47], the eddy viscosity is defined by:

$$\mu_T = \rho k / \omega \tag{2}$$

where μ_T is the velocity vector, k is the turbulence kinetic energy, and ω is the specific dissipation rate. To fulfill the simulation, a simplified two-dimensional model is used in our calculation. The size of the liquid zone was 2 mm × 1 mm, while the frequency of vibration was set as 10 kHz and the displacement of vibration was 5 μm. We employed a microscope (KH-7700, Hirox Inc., Tokyo, Japan) to obtain the radius of the bubble, and the height difference between the center of bubble and the interface of the chip. For a bubble with a radius of 118 µm, the height of the center from the chip was 58 µm, which indicates that the bubble is part spherical. The distribution of the X-velocity of flow can be obtained from the simulation results, as shown in Figure 4a. The maximum absolute value of velocity is approximately 400 µm/s. However, at the zone adjacent to the bubble, the rate of flow was relatively low and approaches zero, which correlates well with the experimental results. Figure 4b indicates that in the bottom area of the liquid, the direction of convective flow was towards the bubble, such that micro-objects and cells could be attracted by the bubble.

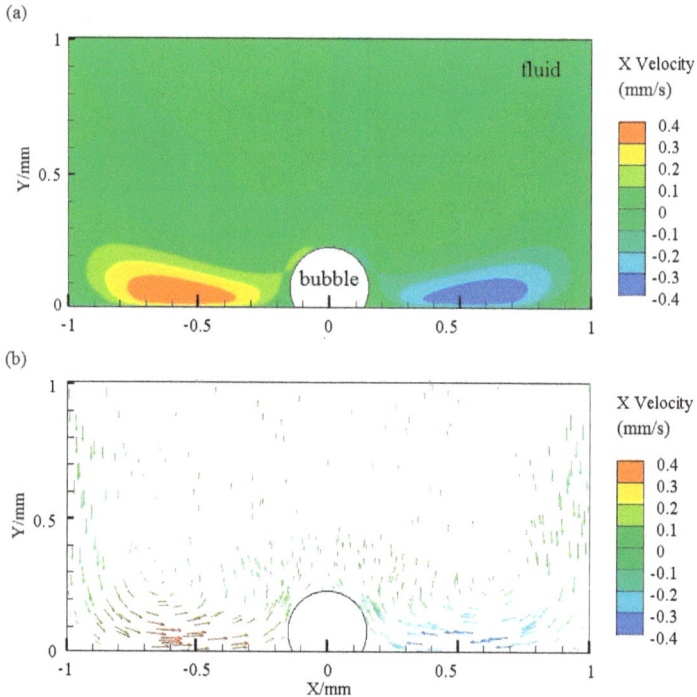

Figure 4. (**a**) Contours of X-velocity of the flow; (**b**) vectors of the flow. The simulation is conducted by ANSYS Fluent software. The diameter of the bubble is 236 µm, the size of the liquid zone is 2 mm × 1 mm, and the frequency and displacement of vibration is 10 kHz and 5 µm, respectively.

3.2. Generation of Bubble

The process of the generation and expansion of the microbubble is shown in Figure 5a–d. When the laser was focused on the liquid-solid interface by the lens, the gold layer absorbed the energy of the light and transferred it into thermal power, and thus the temperature of the liquid near the spot rose. An opto-thermal bubble then generated and expanded continuously because the solubility of the gas generally decreased with increased water temperature. In the initial 5 s, the diameter of the bubble rose quickly. However, the increasing speed gradually then reduced. The diameter of the bubble was determined predominantly by the irradiation time and the power of the laser, as shown in Figure 5e. The greater the power transferred into thermal energy, the more gas separated from the water. Thus, the volume of the bubble is in proportion to the quantity of heat generated. The radius of the bubble is related to the working time and intensity of the laser beam. The growth process of gas in solution can be described by the Lifshitz-Slyozov-Wagner theory [48]:

$$V - V_0 = kIt \tag{3}$$

where V is the volume of the bubble, V_0 is the initial volume of the bubble, I is the power of the laser, k is the efficiency of energy conversion, and t is the irradiation time. The volume of a part-spherical bubble is proportional to the third power of the radius (r) of the bubble:

$$r^3 = r_0^3 + c^{-1}kIt \qquad (4)$$

where r_0 is the initial radius of the bubble, and c is a constant representing the ratio of the volume and radius. As shown in Figure 5f, the experimental results agree to the theoretical curve strongly.

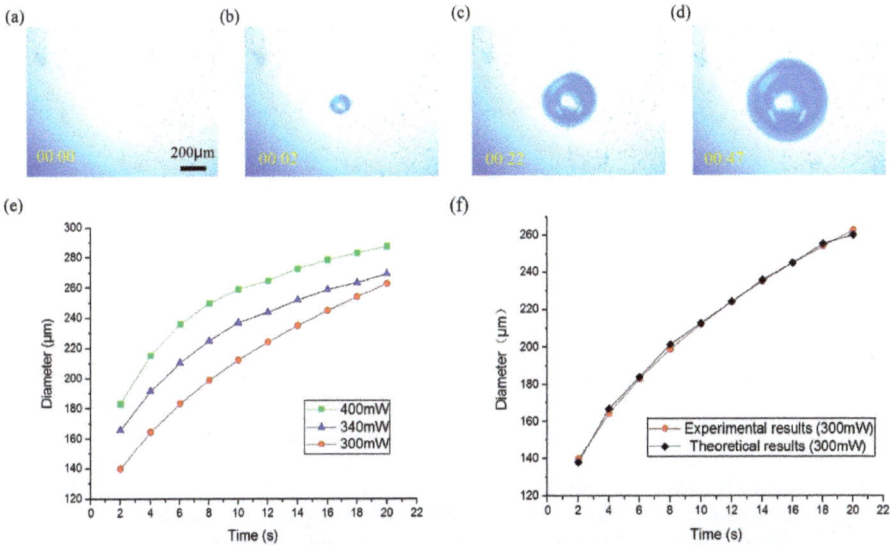

Figure 5. (a–d) Generation and expansion of bubble, the time stamp format is minute:second; (e) growth process of bubble under different conditions where the power of the laser differs; (f) comparison between experimental data and theoretical analysis results, where power is 300 mW.

3.3. Manipulation of Microparticles

To manipulate micro-objects in a liquid reservoir, the function generator and amplifier were turned on to output a sinusoidal voltage, so that the microbubble generated previously was piezo-actuated. The microstreaming around the bubble then attracted objects to the surrounding area progressively, such that micro-objects were collected by the oscillating bubble. Figure 6 demonstrates how the BTG microparticles were captured individually. The diameter of these microballs ranges from 50 to 100 μm. When the bubble is vibrated, a nearby object is trapped initially, and another two balls move towards the bubble at a later stage. Provided the micureobjects were captured, the motion stopped before the streaming became week, as demonstrated by the simulations. Simultaneously, more and more objects were pulled to the bubble continually. The microballs moved quickly, with this manipulation process taking only three seconds.

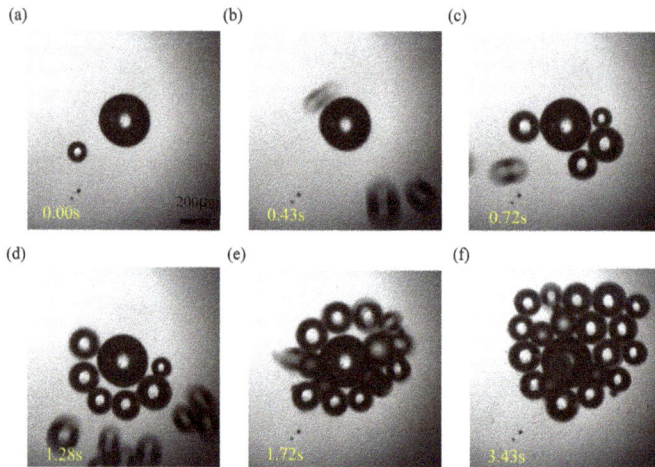

Figure 6. Collection process of microparticles by a bubble; the time format unit is seconds. The frequency and voltage of piezo-actuator is 8 kHz and 30 V; the diameter of bubble and micro glass balls is about 150 μm and 50–100 μm. (**a**) 0.00 s, (**b**) 0.43 s, (**c**) 0.72 s, (**d**) 1.28 s, (**e**) 1.72 s and (**f**) 3.43 s.

To demonstrate the collection ability of the vibrated bubble, the number of micro-objects trapped under different actuation conditions (for example, frequency and amplitude) was studied. Results are shown in Figure 7. In this experiment, there were 30 micro-objects in the liquid reservoir. When the signals were set at 30 V, a bubble could attract more microballs when the vibrating frequency was 4–8 kHz. When the frequency was lower than 500 Hz or higher than 15 kHz, no further objects could be collected. By maintaining a constant frequency of 10 kHz, and gradually increasing the waveform generation gradually, an increasing number of objects were trapped by the bubble. When the voltage was lower than 3 V, the glass balls did not move. Moreover, when the voltage was turned up to 15 V, the ability of the bubble reached its limit. Other conditions, such as the bubble diameter and waveform of the actuation signal, did not affect the number of collected objects.

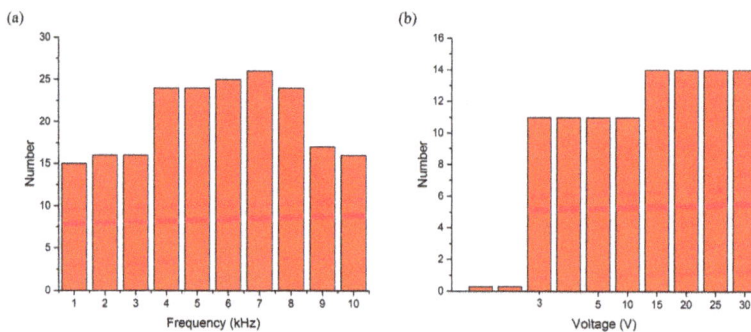

Figure 7. (**a**) Number of micro-objects collected by the bubble at different actuating frequencies; (**b**) number of the microballs collected by the bubble at different actuating voltages.

3.4. Manipulation of Cells

Besides glass balls, the bubbles could also be used to trap bio-objects, including cells. HEK 293 cells are often used in biological and medical experiments, since they are easy to transfer and culture. The process of collecting and manipulating HEK 293 cells using a piezo-actuated bubble is shown in

Figure 8. To manipulate the HEK 293 cells, the opto-thermal bubble was generated in deionized water, as the high viscosity of the culture solution prevented the generation of bubbles. Culture solution and suspended HEK293 cells were then injected into the reservoir on the chip to replace water. HEK 293 cells were usually cultured adherently, and could be suspended in the medium. When the piezo-actuator was turned on, the cells moved towards the oscillated bubble. The diameter of bubble was 70 μm, while the diameter of HEK 293 cells was approximately 20 μm. The two cells in the picture were driven towards the bubble, with their moving distance being about 50 μm, and a speed of 50 μm/s. The frequency and voltage was set at 8 kHz and 30 V, respectively. In our experiments, the activity of the cells was not affected after collection.

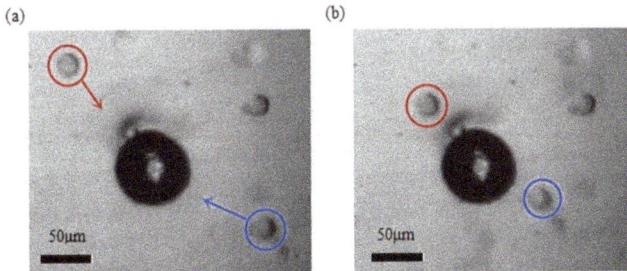

Figure 8. Moving and collecting process of HEK 293 cells by a bubble. (**a**) The cells marked with red and blue circles moves towards the bubble; (**b**) the two cells are collected by the oscillating bubble. The frequency and amplitude of control signal is 8 kHz and 30 V. The diameter of bubble is about 70 μm, and the diameter of the cells is 20 μm. This moving process takes only one second.

In addition, the manipulation method can be used for trapping swimming *pandorina morum* cells. It is a promising application in biology, chemistry, physics, and medicine to manipulate cells with intrinsic motility, as the precise manipulation of the mobile microorganisms (such as bacterial and algal cells) remains difficult [49]. The feasibility and effectiveness of utilizing this technology to manipulate swimming *pandorina morum* cells was investigated, as shown in Figure 9. A bubble was generated on the chip using optically-induced heating. *Pandorina morums* then moved toward the oscillating bubble, and were trapped. Since cells with diameters of 30 μm were smaller than the bubble (whose diameter is more than 100 μm), some of them in the shade of bubble are invisible in the figures. Because the *morum* cells can be self-propelled, this manipulation process took about 5 s to collect these swimming cells far from the bubble, and the moving speed went up to 100 μm/s. The driven voltage used was 8 kHz and 30 V. In our experiments, the activity of *pandorina morum* cells was not affected, and the captured cells could swim away when the bubble was damaged.

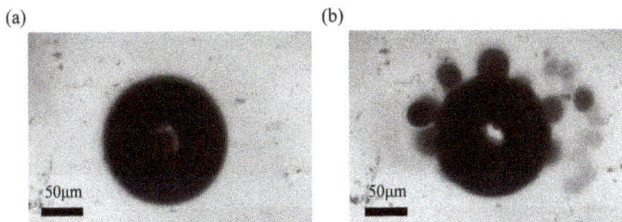

Figure 9. Collection of *pandorina morums* by an oscillated bubble. (**a**) No *morums* around the bubble when the piezo-actuator is off; (**b**) some *morums* are collected by the vibrating bubble. The frequency and amplitude of control signal is 8 kHz and 30 V. The diameter of bubble is about 130 μm, and the diameter of the cells is 30 μm. This collecting process takes five seconds.

3.5. Transportation of Micro-Objects

In addition, we could collect micro-objects at different destinations by changing the position of the bubble. Because the mobile stage was connected with a laser and lens, we could generate bubbles at arbitrary locations, so that the transportation of microballs was complete. When a bubble burst or was damaged, the micro-objects could be re-collected at a different destination where a new bubble was generated. Figure 10 illustrates the procedures of the collection and transportation of the micro-objects. The diameter of bubble in Figure 10b–d is more than 300 μm, while the diameter of bubble in Figure 10f–h is about 200 μm, and the diameter of mciro balls is 50 to 100 μm. The moving distance of the balls was more than 2 mm in the chip. The frequency and amplitude of the input signal of the piezo-actuator was 8 kHz and 30 V, respectively. The first bubble was generated on the chip and the particles were captured. Since a bubble disappeared at its natural frequency, it could be damaged by changing the frequency of the piezoelectric stack, resulting in the collected objects redispersing in the aqueous media. In addition, if we turned on the laser again, the bubble grew continuously to its limit and broke. If we moved the laser spot to a new destination 1 mm away from the original location and the second bubble is created, the microballs collected around this new oscillatory bubble. This manipulation process was repeatable and micro-objects could be transported continuously. The working distance of the vibrated bubble could cover the whole chip. Thus, the microballs can be transported by the bubble over considerable distances.

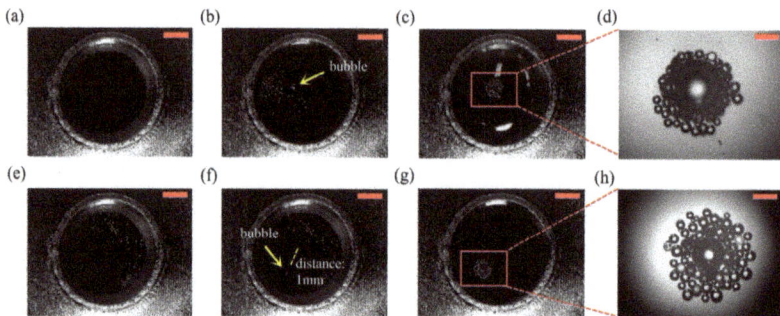

Figure 10. Collection and transportation of micro-objects: (**a**) liquid reservoir; (**b**) first bubble generates on the chip; (**c,d**) vibrated bubble attracts microballs—this manipulation takes about 3 s; (**e**) objects spread after the first bubble is damaged; (**f**) a new bubble appears at new location 1 mm far from the original position; (**g,h**) microballs are re-collected and transported by the oscillated bubble—the collection process takes 4 s. Scale bars in (**a–c**), (**e–g**) are 1 mm, and are 200 μm in (**d,h**).

4. Conclusions

A novel micromanipulation and transportation technology using opto-thermally generated and piezo-actuated bubbles is proposed in this paper. Although optothermal generation and acoustic oscillation of microbubbles have been well studied and applied to micromanipulator and lab-on-a-chip, using these techniques in one chip has not been realized so far. The ability of the bubble to capture objects, which is demonstrated by the computational fluid dynamics simulation, is related to characteristics of the control signal. Since the bubble can be generated at arbitrary locations in the open chip environment, the microballs and cells can be collected and transported efficiently as a mass. Manipulation in open chip environment, along with the function of collection and transportation of micro-objects, make it a promising micromanipulation method. The method based on vibrated bubbles is expected to cooperate with other scientific instruments, such as AFM and SICM, under the open operating condition. The advantages of this manipulation technique, such as versatility and simplicity, makes it a good candidate for actuation of self-sufficient, stand-alone microfluidic systems [50], if the laser and acoustic wave sources can be miniaturized further.

Acknowledgments: This Research is supported by the National Natural Science Foundation of China (Grant No. 61573339) and the CAS/SAFEA International Partnership Program for Creative Research Teams.

Author Contributions: L.D. and N.J. conceived and designed the experiments; L.D. and X.W. fabricated the devices and performed the experiments; L.D. analyzed the data; L.D. and N.J. wrote the paper; L.L. supervised the entire work.

Conflicts of Interest: The authors declare no conflict of interest.

References

1. Block, S.M.; Blair, D.F.; Berg, H.C. Compliance of bacterial optical polyhooks measured with tweezers. *Nature* **1989**, *338*, 514–518. [CrossRef] [PubMed]
2. Kuo, S.C.; Sheetz, M.P. Force of single kinesin molecules measured with optical tweezers. *Science* **1993**, *260*, 232–234. [CrossRef] [PubMed]
3. Grigorenko, A.N.; Roberts, N.W.; Dickinson, M.R.; Zhang, Y. Nanometric optical tweezers based on nanostructured substrates. *Nat. Photonics* **2008**, *2*, 365–370. [CrossRef]
4. Liesener, J.; Reicherter, M.; Haist, T.; Tiziani, H.J. Multi-functional optical tweezers using computer-generated holograms. *Opt. Commun.* **2000**, *185*, 77–82. [CrossRef]
5. Huang, S.B.; Wu, M.H.; Lin, Y.H.; Hsieh, C.H.; Yang, C.L.; Lin, H.C.; Tseng, C.P.; Lee, G.B. High-purity and label-free isolation of circulating tumor cells (CTCs) in a microfluidic platform by using optically-induced-dielectrophoretic (ODEP) force. *Lab Chip* **2013**, *13*, 1371–1383. [CrossRef] [PubMed]
6. Liang, W.F.; Liu, N.; Dong, Z.L.; Liu, L.Q.; Mai, J.D.; Lee, G.B.; Li, W.J. Simultaneous separation and concentration of micro- and nano-particles by optically induced electrokinetics. *Sens. Actuators A Phys.* **2013**, *193*, 103–111. [CrossRef]
7. Lin, Y.H.; Lee, G.B. Optically induced flow cytometry for continuous microparticle counting and sorting. *Biosens. Bioelectron.* **2008**, *24*, 572–578. [CrossRef] [PubMed]
8. Schmid, L.; Wixforth, A.; Weitz, D.A.; Franke, T. Novel surface acoustic wave (SAW)-driven closed PDMS flow chamber. *Microfluid. Nanofluid.* **2012**, *12*, 229–235. [CrossRef]
9. Shilton, R.J.; Travagliati, M.; Beltram, F.; Cecchini, M. Nanoliter-droplet acoustic streaming via ultra highfrequency surface acoustic waves. *Adv. Mater.* **2014**, *26*, 4941–4946. [CrossRef] [PubMed]
10. Evander, M.; Nilsson, J. Acoustofluidics 20: Applications in acoustic trapping. *Lab Chip* **2012**, *12*, 4667–4676. [CrossRef] [PubMed]
11. Ding, X.Y.; Peng, Z.L.; Lin, S.C.S.; Geri, M.; Li, S.X.; Li, P.; Chen, Y.C.; Dao, M.; Suresh, S.; Huang, T.J. Cell separation using tilted-angle standing surface acoustic waves. *Proc. Natl. Acad. Sci. USA* **2014**, *111*, 12992–12997. [CrossRef] [PubMed]
12. Li, Z.; Abbott, J.J.; Dong, L.; Kratochvil, B.E.; Bell, D.; Nelson, B.J. Artificial bacterial flagella: Fabrication and magnetic control. *Appl. Phys. Lett.* **2009**, *94*, 064107.
13. Huang, H.W.; Sakar, M.S.; Petruska, A.J.; Pane, S.; Nelson, B.J. Soft micromachines with programmable motility and morphology. *Nat. Commun.* **2016**, *7*, 12263. [CrossRef] [PubMed]
14. Nelson, B.J.; Kaliakatsos, I.K.; Abbott, J.J. Microrobots for Minimally Invasive Medicine. *Annu. Rev. Biomed. Eng.* **2010**, *12*, 55–85. [CrossRef] [PubMed]
15. Wang, J.Y.; Jiao, N.D.; Tung, S.; Liu, L.Q. Automatic path tracking and target manipulation of a magnetic microrobot. *Micromachines* **2016**, *7*, 212. [CrossRef]
16. Chiu, S.H.; Liu, C.H. An air-bubble-actuated micropump for on-chip blood transportation. *Lab Chip* **2009**, *9*, 1524–1533. [CrossRef] [PubMed]
17. Dijkink, R.; Ohl, C.D. Laser-induced cavitation based micropump. *Lab Chip* **2008**, *8*, 1676–1681. [CrossRef] [PubMed]
18. Wu, T.H.; Chen, Y.; Park, S.Y.; Hong, J.; Teslaa, T.; Zhong, J.F.; Di Carlo, D.; Teitell, M.A.; Chiou, P.Y. Pulsed laser triggered high speed microfluidic fluorescence activated cell sorter. *Lab Chip* **2012**, *12*, 1378–1383. [CrossRef] [PubMed]

19. Tan, W.H.; Takeuchi, S. A trap-and-release integrated microfluidic system for dynamic microarray applications. *Proc. Natl. Acad. Sci. USA* **2007**, *104*, 1146–1151. [CrossRef] [PubMed]
20. Ahmed, D.; Mao, X.L.; Shi, J.J.; Juluri, B.K.; Huang, T.J. A millisecond micromixer via single-bubble-based acoustic streaming. *Lab Chip* **2009**, *9*, 2738–2741. [CrossRef] [PubMed]
21. Hellman, A.N.; Rau, K.R.; Yoon, H.H.; Bae, S.; Palmer, J.F.; Phillips, K.S.; Allbritton, N.L.; Venugopalan, V. Laser-induced mixing in microfluidic channels. *Anal. Chem.* **2007**, *79*, 4484–4492. [CrossRef] [PubMed]
22. Wijngaart, W.; Chugh, D.; Man, E.; Melin, J.; Stemme, G. A low-temperature thermopneumatic actuation principle for gas bubble microvalves. *J. Microelectromech. Syst.* **2007**, *16*, 765–774. [CrossRef]
23. Xu, Y.C.; Lv, Y.; Wang, L.; Xing, W.L.; Cheng, J. A microfluidic device with passive air-bubble valves for real-time measurement of dose-dependent drug cytotoxicity through impedance sensing. *Biosens. Bioelectron.* **2012**, *32*, 300–304. [CrossRef] [PubMed]
24. Prakash, M.; Gershenfeld, N. Microfluidic bubble logic. *Science* **2007**, *315*, 832–835. [CrossRef] [PubMed]
25. Zhang, K.; Jian, A.; Zhang, X.; Wang, Y.; Li, Z.; Tam, H.-Y. Laser-induced thermal bubbles for microfluidic applications. *Lab Chip* **2011**, *11*, 1389–1395. [CrossRef] [PubMed]
26. Marmottant, P.; Raven, J.P.; Gardeniers, H.; Bomer, J.G.; Hilgenfeldt, S. Microfluidics with ultrasound-driven bubbles. *J. Fluid. Mech.* **2006**, *568*, 109–118. [CrossRef]
27. Khoshmanesh, K.; Almansouri, A.; Albloushi, H.; Yi, P.; Soffe, R.; Kalantar-Zadeh, K. A multi-functional bubble-based microfluidic system. *Sci. Rep.* **2015**, *5*, 9942. [CrossRef] [PubMed]
28. Hashmi, A.; Yu, G.; Reilly-Collette, M.; Heiman, G.; Xu, J. Oscillating bubbles: A versatile tool for lab on a chip applications. *Lab Chip* **2012**, *12*, 4216–4227. [CrossRef] [PubMed]
29. Marmottant, P.; Hilgenfeldt, S. Controlled vesicle deformation and lysis by single oscillating bubbles. *Nature* **2003**, *422*, 153–156. [CrossRef] [PubMed]
30. Rogers, P.; Neild, A. Selective particle trapping using an oscillating microbubble. *Lab Chip* **2011**, *11*, 3710–3715. [CrossRef] [PubMed]
31. Sang, K.C.; Sung, K.C. On-chip manipulation of objects using mobile oscillating bubbles. *J. Micromech. Microeng.* **2008**, *18*, 125024.
32. Zhao, C.L.; Xie, Y.L.; Mao, Z.M.; Zhao, Y.H.; Rufo, J.; Yang, S.K.; Guo, F.; Mai, J.D.; Huang, T.J. Theory and experiment on particle trapping and manipulation via optothermally generated bubbles. *Lab Chip* **2014**, *14*, 384–391. [CrossRef] [PubMed]
33. Sang, K.C.; Kyehan, R.; Sung, K.C. Bubble actuation by electrowetting-on-dielectric (EWOD) and its applications: A review. *Int. J. Precis. Eng. Manuf.* **2010**, *11*, 991–1006.
34. Lee, J.H.; Lee, K.H.; Chae, J.B.; Rhee, K.; Chung, S.K. On-chip micromanipulation by AC-EWOD driven twin bubbles. *Sens. Actuators A Phys.* **2013**, *195*, 167–174. [CrossRef]
35. Lee, K.H.; Lee, J.H.; Won, J.M.; Rhee, K.; Chung, S.K. Micromanipulation using cavitational microstreaming generated by acoustically oscillating twin bubbles. *Sens. Actuators A Phys.* **2012**, *188*, 442–449. [CrossRef]
36. Hu, W.; Ishii, K.S.; Ohta, A.T. Micro-assembly using optically controlled bubble microrobots. *Appl. Phys. Lett.* **2011**, *99*, 094103. [CrossRef]
37. Hu, W.Q.; Ishii, K.S.; Fan, Q.H.; Ohta, A.T. Hydrogel microrobots actuated by optically generated vapour bubbles. *Lab Chip* **2012**, *12*, 3821–3826. [CrossRef] [PubMed]
38. Hu, W.Q.; Fan, Q.H.; Ohta, A.T. An opto-thermocapillary cell micromanipulator. *Lab Chip* **2013**, *13*, 2285–2291. [CrossRef] [PubMed]
39. Xie, Y.L.; Nama, N.; Li, P.; Mao, Z.M.; Huang, P.H.; Zhao, C.L.; Costanzo, F.; Huang, T.J. Probing cell deformability via acoustically actuated bubbles. *Small* **2016**, *12*, 902–910. [CrossRef] [PubMed]
40. Dai, L.G.; Jiao, N.D.; Liu, L.Q. Particle Manipulation via Opto-thermally Generated Bubbles in Open Chip Environment. In Proceedings of the 16th International Conference on Nanotechnology (IEEE-Nano), Sendai, Japan, 22–25 August 2016; pp. 30–33.
41. Chen, Y.; Lee, S. Manipulation of Biological Objects Using Acoustic Bubbles: A Review. *Integr. Comp. Biol.* **2014**, *54*, 959–968. [CrossRef] [PubMed]
42. Duffy, D.C.; McDonald, J.C.; Schueller, O.J.A.; Whitesides, G.M. Rapid prototyping of microfluidic systems in poly(dimethylsiloxane). *Anal. Chem.* **1998**, *70*, 4974–4984. [CrossRef] [PubMed]
43. McDonald, J.C.; Whitesides, G.M. Poly(dimethylsiloxane) as a material for fabricating microfluidic devices. *Accounts Chem. Res.* **2002**, *35*, 491–499. [CrossRef]

44. Frisch, U.; Hasslacher, B.; Pomeau, Y. Lattice-gas automata for the Navier-Stokes equation. *Phys. Rev. Lett.* **1986**, *56*, 1505–1508. [CrossRef] [PubMed]
45. Wilcox, D.C. Reassessment of the scale-determining equation for advanced turbulence models. *AIAA J.* **1988**, *26*, 1299–1310. [CrossRef]
46. Piomelli, U.; Zang, T.A.; Speziale, C.G.; Hussaini, M.Y. On the large-eddy simulation of transitional wall-bounded flows. *Phys. Fluids* **1990**, *2*, 257–265. [CrossRef]
47. Wilcox, D.C. Simulation of transition with a two-equation turbulence model. *AIAA J.* **1994**, *32*, 247–255. [CrossRef]
48. Lifshitz, I.M.; Slyozov, V.V. The kinetics of precipitation from supersaturated solid solutions. *J. Phys. Chem. Solids* **1961**, *19*, 35–50. [CrossRef]
49. Xie, S.X.; Jiao, N.D.; Tung, S.; Liu, L.Q. Controlled regular locomotion of algae cell microrobots. *Biomed. Microdevices* **2016**, *18*, 47. [CrossRef] [PubMed]
50. Boyd-Moss, M.; Baratchi, S.; Di Venere, M.; Khoshmanesh, K. Self-contained microfluidic systems: A review. *Lab Chip* **2016**, *16*, 3177–3192. [CrossRef] [PubMed]

micromachines

MDPI

Article

Development of the Electric Equivalent Model for the Cytoplasmic Microinjection of Small Adherent Cells

Florence Hiu Ling Chan [1,3], **Runhuai Yang** [2] and **King Wai Chiu Lai** [1,3,*]

1   Department of Mechanical and Biomedical Engineering, City University of Hong Kong, Hong Kong, China; florencehlchan@gmail.com
2   Department of Biomedical Engineering, Anhui Medical University, Hefei 230032, China; yrunhuai@foxmail.com
3   Centre for Robotics and Automation, City University of Hong Kong, Hong Kong, China
*   Correspondence: kinglai@cityu.edu.hk; Tel.: +852-3442-9099

Received: 2 May 2017; Accepted: 4 July 2017; Published: 8 July 2017

Abstract: A novel approach utilizing current feedback for the cytoplasmic microinjection of biological cells is proposed. In order to realize the cytoplasmic microinjection on small adherent cells (diameter < 30 μm and thickness < 10 μm), an electrical model is built and analyzed according to the electrochemical properties of target cells. In this study, we have verified the effectiveness of the current measurement for monitoring the injection process and the study of ion channel activities for verifying the cell viability of the cells after the microinjection.

Keywords: cell cytoplasmic microinjection; small adherent cells; equivalent-circuit model

1. Introduction

Cell injection is a promising micromanipulation method in the biological field [1]. Efficient tools for precisely injecting drugs into cells are of significant value to the development of novel therapeutics [2], and are also important for the study of cellular activity in biological fields [3]. Progresses in biological engineering such as drug screening and transfection require the precise manipulation of single biological cells [4]. Several methods, including drug delivery by either nano-vehicles or nanoparticles, may be used to penetrate the cell membrane and perform intracellular delivery [5,6]. In addition, viral vector-based methods are very effective for injecting the target nuclide acid into various kinds of cells, ranging from the ex vivo transduction of hematopoietic cells to in vivo transgene expression through the optimization of tissue-specific cells. However, safety remains a great concern for gene therapies using the viral vector [7]. Besides, a nano-needle is very efficient for delivering different molecules and materials including antibodies, quantum dots, and nanoparticles to primary hippocalmpus neuron cells and NIH3T3 fibroblast cells, which both have a diameter of around 100 μm [8]. Unfortunately, the injection performance on small adherent cells using the nano-needles is unknown at present.

A micro-pipette-based microinjection is a process for the delivery of target materials by penetrating living cells using a micropipette. It has been proven to work effectively and has the advantages of a precise control of the delivery dosage and a high transduction efficiency [9]. However, the drawback of pipette-based methods is the need for a well-trained operator, thus requiring a lot of time and money [10]. This led to the demand of an automatic pipette-based cell microinjection in order to increase the success rate and reduce human errors [11]. Several automated cell injection approaches with different kinds of feedbacks were reported, including force feedback, impedance feedback, and visual feedback. The microNewton force feedback approach was reported to largely improve the work efficiency and success rate of microRNA and DNA injections to suspended mouse embryos (around 100 μm in diameter) and zebrafish cells (1.3 mm in diameter) [12]. For smaller cells, the penetration force analysis of several cell lines, which are L929, HeLa, 4T1, and TA3 HA II cells, was demonstrated using atomic force microscopy (AFM) [2], which helps the researcher to understand the cell mechanics. Impedance feedback for a robot-assisted cell microinjection system performed on zebrafish fish embryos was introduced to enhance the accuracy of the system feedback [2,13]. A vision-based force measurement for cell microinjection was performed on mouse oocytes (around 100 μm in diameter) and drosophila embryos (around 500 μm in diameter) [14]. A visual-based system with impedance control for cell injection was reported to perform successfully on Zebrafish embryos [15]. Some researchers suggested that the visual feedback may be a possible solution for the micro-injection of small cells with a diameter of less than 100 μm [16]. However, another challenge which arises from visual feedback is the difficulty in observing the contact between the tip and the cell to determine the penetration status. In recent years, more genetic research has driven the demand of an automated cell microinjection system, and some in vitro cell microinjection systems were reported in a review [17]. All of the microinjection systems mentioned are for suspending cells in an oval shape. Nevertheless, some typical human cells are adherent and small, such as neuroblastoma cells and human epithelia cells. Recently, an automated injection system for suspending small cells (diameter < 30 μm) was reported, which was operated with a microfluidic cell holder chip [11,18]. Small adherent cells are of importance in cell biology because of the size of human cells, which usually ranges from 7 μm to 25 μm, and also cell transfection using human cells with the precise delivery of target materials is challenging to operate, but there is great demand for it to be automated [11]. In the past, there have been quite a number of researches about the automated injection of large and round cells such as the zebra fish embryo [1,4,9,19,20]. In contrast, there are only a few number of researches talking about the automated injection of small adherent cells [21]. For instance, SHSY-5Y cells (human neuroblastoma cells) are used frequently for studying Alzheimer's disease. HEK-293 cells (human embryonic kidney cells) are used frequently for studying gene therapy. However, it is more difficult to implement either the manual or automated injection on human cells due to their small size.

To overcome the challenges, a current feedback approach is proposed to work for automating the micro-injection of the neuronal cell line (<30 μm). In this study, an equivalent electrical model was established to monitor two situations. The real time current response during the microinjection was studied for the effectiveness of the injection. In addition, the current trace under voltage stimulus was recorded in order to verify the viability of the cells after microinjection.

2. The Electric Equivalent Model for Cytoplasmic Microinjection and Cell Viability Verification

In order to monitor the electrical signal of biological cells for microinjection, an electric equivalent model of a cell cytoplasmic microinjection with a micro-pipette has been proposed, as illustrated in Figure 1. A cell interacting with a micro-pipette can be regarded as an electrical circuit containing several electrical components. E_{Na}, E_K, and $E_{Other\ ions}$ are the equilibrium (Nernst) potentials of Na^+, K^+, and other ions between the cell membrane. In normal condition, E_K and E_{Na} are about -90 mV and $+60$ mV [22] due to the concentration difference between the outer and inner fluid of living cells. Naturally, the external fluid usually contains 0.44 M of Na+, and the internal fluid, which is the cytoplasm, usually contains 0.5 M of K^+ and a number of unspecified anions such as phosphates, amino acids, and negatively charged proteins [23]. In this study, cells were bathed in the buffer solution as the outer fluid of living cells, called extracellular solution. Meanwhile, another solution was prepared as the inner fluid of living cells, called intracellular solution.

The capacitance (C_M) of the cell membrane exists because of the different concentration of total ions between the outer and inner fluid. The glass layer structure of the pipette acts as a capacitor, $C_{Pipette}$, which gives the capacitance transient curve when the micropipette is immersed in the extracellular solution. The capacitance transient curve can be neutralized by injecting the opposite capacitance current via adjustment with observation of the electrical signal. R_{Na}, R_K, and $R_{Other\ ions}$ are the variable resistances of the ionic currents contributed by Na^+, K^+, and other ions, respectively. Their resistances vary because the various ions (Na^+, K^+, and other ions) pass through the cell members when the corresponding ion channels are activated under a different stimulus [24]. R_{Seal} is a resistance due to the formation of a tight and strong seal between the cell surface and the pipette before the microinjection. Access resistance (R_{Access}) results from the geometry of the pipette plus any obstruction at the tip by the adherent membrane [19].

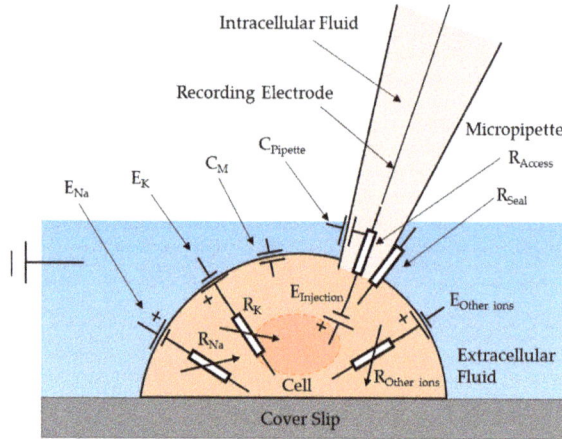

Figure 1. The electric equivalent model of the cell ctoplasmic injection.

In the model, $E_{Injection}$ is hypothesized to exist when the intracellular solution is injected into a cell using a micropipette. It is believed that the injection will cause a significant change of the membrane potential because of the change of ion concentrations across the membrane. During the injection, the injected intracellular solution immediately mixes with the cell cytoplasm leading to a change of ion concentration in the cytoplasm, while the concentration of the extracellular solution remains unchanged. $E_{Injection}$ can be expressed as shown in Equation (1). $E_{Cell\ Membrane}$ can be calculated by the Nernst Equation, as shown in Equation (2):

$$E_{Injection} = New\ E_{Cell\ Membrane} - Old\ E_{Cell\ Membrane} \tag{1}$$

where *New* $E_{Cell\ Membrane}$ is the membrane potential after the injection, and *Old* $E_{Cell\ Membrance}$ is the membrane potential before the injection:

$$E_{Cell\ Membrane} = \sum_{j=0}^{n} \frac{RT}{ZF} In \frac{[ion_j]_o}{[ion_j]_i} \tag{2}$$

where R is the ideal gas constant, $R = 8.314472\ J\ K^{-1}\cdot mol^{-1}$, T is the temperature in kelvins, F is Faraday's constant (coulombs per mole), $F = 9.648533 \times 10^4\ C\cdot mol^{-1}$, $[ion]_o$ is the extracellular concentration of that particular ion (in moles per cubic meter), $[ion]_i$ is the intracellular concentration of that particular ion (in moles per cubic meter), and Z is the number of moles of electrons transferred in the cell reaction or half-reaction.

By taking into consideration all of the ions of the intracellular solution, as mentioned in Section 3, $E_{Injection}$ can be expressed as shown in Equation (3).

$$E_{cell\ membrane} = \frac{RT}{(+1)F} In \frac{[Na^+]_o}{[Na^+]_i} + \frac{RT}{(+1)F} In \frac{[K^+]_o}{[K^+]_i} + \frac{RT}{(+2)F} In \frac{[Mg^{2+}]_o}{[Mg^{2+}]_i} + \frac{RT}{(-2)F} In \frac{[Cl^{2-}]_o}{[Cl^{2-}]_i}$$
$$+ \frac{RT}{(Z)F} In \frac{[Other\ ions]_o}{[Other\ ions]_i} \tag{3}$$

where $\frac{RT}{(Z)F} In \frac{[Other\ ions]_o}{[Other\ ions]_i} \cong 0$, since the concentration of ion is very small.

Since the major injected ions are K$^+$ and Na$^+$, the equation of cell membrane potential can be simplified in Equation (4).

$$E_{cell\ membrane} = \frac{RT}{(+1)F} ln \frac{[Na^+]_o}{[Na^+]_i} + \frac{RT}{(+1)F} ln \frac{[K^+]_o}{[K^+]_i} \tag{4}$$

To study and verify this electrical model, the patch clamp technique was used with a low noise operation amplifier that can apply voltage stimulus for controlling the voltage potential of the cell membrane, called the membrane potential (V_m). It can also acquire the total current response of the cell membrane, called the membrane current (I_m).

When a cell is connected to the operational amplifier, the electric equivalent circuit of the cell cytoplasmic microinjection is constructed as shown in Figure 2. The operation amplifier acts as a voltage source (E_{OpAmp}) in series with a resistor (R_{OpAmp}) in order to apply the voltage stimulus which can evolve the ion flow along the ion channels. There are five switches in the circuit, including the switch (S_{seal}) for the sealing resistance and the switch ($S_{Injection}$) for the microinjection of the cell. In this circuit, S_{seal} is always open as R_{Seal} is extremely large compared with other resistances, and hence the connection to R_{Seal} will act as an open circuit. The function of the switch ($S_{Injection}$) for the microinjection of the cell is to connect the $E_{Injection}$ when the intracellular solution is injected into the cell. S_K, S_{Na}, and $S_{Other\ ions}$ act like the switches to connect R_{Na}, R_K, and $R_{Other\ ions}$, which will be the maximum when there is the least ion channel activity, and vice versa.

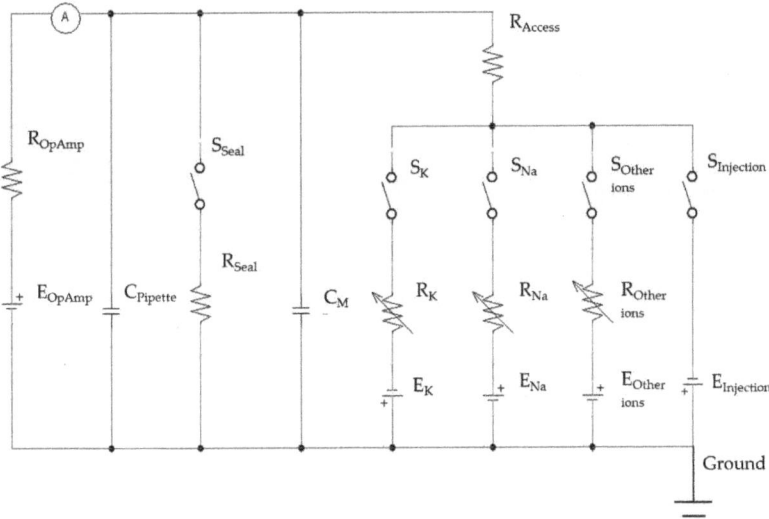

Figure 2. The equivalent circuit of the current feedback model for the cell cytoplasmic microinjection and cell viability verification.

In the measurement, the membrane current (I_M) can be recorded. According to the circuit model, I_M is composed of several components, as shown in Equation (5), including the current caused by the pipette capacitance ($I_{Pipette}$), the current passing through the sealing resistance (I_{Seal}), the current passing through the access resistor (I_{Access}), and the current caused by the membrane capacitance (I_c). I_C is produced by the charge accumulation at the outer and inner membrane surface when V_m changes.

$$I_m = I_{pipette} + I_{Seal} + I_C + I_{Access} \tag{5}$$

To simplify Equation (5), $I_{Pipette}$ can be cancelled out using a setting of the patch clamp. In addition, I_{Seal} is usually small enough to be neglected when R_{Seal} is very large. I_C is nonzero only when V_m is changing [23]. Throughout the microinjection, V_m is controlled at a constant level. When studying the ion channel activities, the stimulus of a square voltage pulse is applied to the cell. Hence, I_C will be nearly zero because V_m is always constant and V_m only changes at the brief instants when the voltage is stepped to a new value. In other word, $I_C = C_m \frac{dV}{dt} \cong 0$ under the stimulus of a square voltage pulse. I_{Access} is composed of the current passing through the sodium, potassium, and other ion channels (I_{Na}, I_K and $I_{Other\ ions}$) [25], as well as the current drop due to the injection ($I_{Injection}$) in Equation (6).

$$I_{Access} = I_{Na} + I_K + I_{Other\ ions} + I_{Injection} \tag{6}$$

By combining Equations (5) and (6), the membrane current (I_M) will be directly equal to I_{Access}, as shown in Equation (7).

$$I_m = I_{Access} = I_{Na} + I_K + I_{Other\ ions} + I_{Injection} \tag{7}$$

Therefore, the electric equivalent circuit of the cell cytoplasmic microinjection can be simplified as shown in Figure 3.

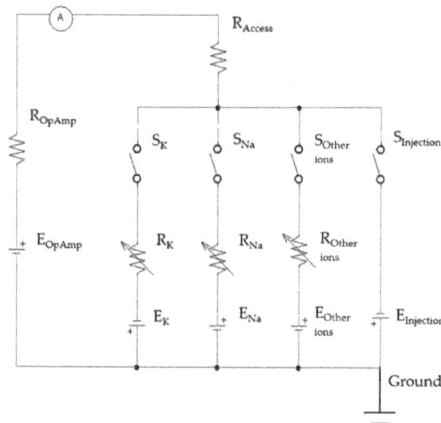

Figure 3. The simplified circuit of the current feedback model for the cell cytoplasmic microinjection and cell viability verification.

2.1. The Equivalent Model for the Cell Cytoplasmic Microinjection

The microinjection was performed on a single living cell when it was at the resting state. S_K, S_{Na}, and $S_{Other\ ions}$ are anticipated to be open because R_{Na}, R_K, and $R_{Other\ ions}$ are expected to be very large since nearly none of the ion channels will be activated without the voltage stimulus. Hence, I_{Na}, I_K, and $I_{Other\ ions}$ will become zero. Subsequently, I_M will only be equal to the current drop, as R_{Access} is connected to $E_{Injection}$, induced by the concentration difference of the ions in the cytoplasm of the living cell due to the microinjection of the intracellular solution, as shown in Equation (8).

$$I_m = I_{Injection} = \frac{E_{Injection}}{R_{Access}} \tag{8}$$

2.2. The Electric Model for Cell Viability Verification

After the injection to a living cell, its vitality and viability are of great concern and can be verified by the study of ion channel activities. When the cells are alive after the microinjection, the activities of the ion channels can be activated by applying the voltage stimulus. Under the voltage stimulus, the cell membrane current I_M of the normal cell can be calculated by Equation (9). In this situation, R_K, R_{Na}, and $R_{Other\ ions}$ are very low, so those ions can easily pass through the cell membrane when the Na$^+$, K$^+$, and other ion channels are activated. Moreover, $I_{Other\ ions}$ can be neglected when compared with I_{Na} and I_K because the activities of the Na$^+$ and K$^+$ ion channels are dominant. Therefore, the I_M of a normal cell is equal to the sum of I_{Na} and I_K, as shown in Equation (10).

$$I_m = I_{Na} + I_K + I_{Other\ ions} \tag{9}$$

where $I_{Other\ ions} \cong 0$.

$$I_m = I_{Na} + I_K \tag{10}$$

where the current response of I_K and I_{Na} depends on R_K and R_{Na}, respectively, as shown in Equations (11) and (12).

$$I_K = \frac{V_m - E_K}{R_K} \tag{11}$$

$$I_{Na} = \frac{V_m - E_{Na}}{R_{Na}} \tag{12}$$

R_K and R_{Na} are variables based on the number of activated ion channels for passing Na$^+$ and K$^+$ ions across the membrane.

3. The Experimental Setup of Cell Cytoplasmic Microinjection

The experimental setup consists of several units (as shown in Figure 4), which are an extracellular solution to bath the cell in order to provide cells with an in vitro environment like in vivo, an intracellular solution to act as an electrolyte connected to a low-noise amplifier and to be injected inside the cell during the microinjection process, a low-noise amplifier (Model: Axopatch 200B from Molecular Device, Sunnyvale, CA, USA) to amplify the recording current signal from the cell, a data digitalizer (Digidata 1440A from Molecular Devices) and computer for generating the voltage stimulus waveform and data acquisition of the current signal from the cell membrane, and the inverted microscope (Model: Eclipse Ti from Nikon, Tokyo, Japan) to monitor the cell cytoplasmic microinjection. The adherent cells grown on cover slips were prepared. In each experiment, the cover slip with adherent cells was placed inside a petri dish, filled with extracellular solution. When implementing the microinjection, the force applied on the cell via the micropipette was vertical, which would not lead to the sliding of the cover slip and movement of the cell. In addition, a Faraday cage is used to reduce the background signal in the surrounding area since the patch clamp recording is very sensitive to background noise.

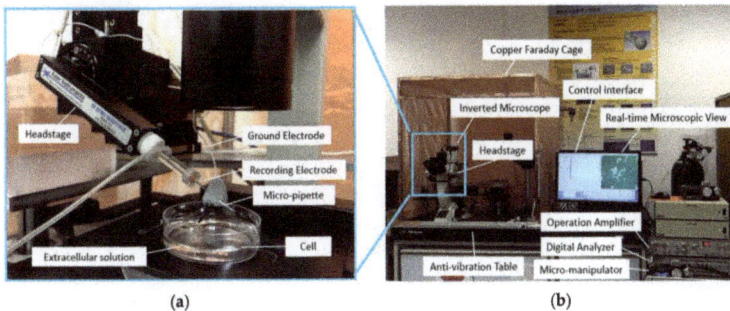

Figure 4. The experimental setup of the current feedback injection: (**a**) the close-up view; (**b**) the overal view.

3.1. The Cell Cytoplasmic Microinjection Process

In this process, the cell cytoplasmic micro-injection process was divided into five main steps. Firstly, a micropipette was made from borosilicate glass capillaries (BF120-69-7.5 from Sutter Instrument, Novato, CA, USA). Each glass capillary was pulled by a CO_2 laser-based micro-pipette puller (Model: P-2000 from Sutter Instrument). As a result, the glass capillaries were divided into two halves, which were the micropipettes with the tip size of 1 μm. From the literature, the common tip size ranges from 0.05 μm to 1 μm [11,17,19,22]. Secondly, SHSY-5Y cells were weekly prepared by cell passage and cell culture. Thirdly, the extracellular solution and the intracellular solution were prepared with various ions. The details of the cell culture and the solution preparation will be shown in the coming sections. Fourthly, the injection process was initiated by positioning the micro-pipette filled with intracellular solution to approach the cell surface using the micro-manipulator. The electrode placed inside the micropipette was connected to the headstage and hence to the operational amplifier. A test square voltage pulse of 20 mV for 10 ms was applied to the cell from the operational amplifier.

When the pipette was immersed into the solution, a typical positive pressure was applied using a 10 mL syringe by using a displacing plunger of about 1 mL to remove any contaminations at the tip. The resistance value was around 4–8 MΩ at the bath position. Afterwards, the pipette tip slowly approached the cell surface until there was an obvious increase of 3 to 5 MΩ in the total resistance and a significant decrease in the test pulse amplitude. When the current signal became steady, the positive pressure could be released rapidly. By applying suction through retracting air using a 5 mL syringe, the plunger was retracted at 0.1 mL/s for a second. Consequently, a very tight seal between the pipette and cell surface can be created with a GΩ formation where the total resistance of the electrical model will be around 1 GΩ. Afterwards, the voltage potential of the cell membrane was controlled at −70 mV and a stronger pressure was applied to the micropipette through retracting air using a 5 mL syringe. The plunger was retracted at 1 mL/s until a significant change of the current response was observed with the existence of capacitive current trace from the cell membrane. In the circuit, the membrane current was measured spontaneously and continuously while the voltage potential was kept at −70 mV throughout the microinjection process. Before starting the microinjection experiment, the baseline of the electric current response was always adjusted to zero. Consequently, the current due to the microinjection can be acquired without the distortion from the original ion activities.

During the microinjection, the current signal of the cell membrane was monitored. The permanent drop of current level was consistently observed right after each injection. In each injection, the injection volume was estimated to be 19 pL. The estimation was based on recording the volume of the cell cytoplasm before and after the injection using optical images. In another study, the injection volume for cells (25 μm) was measured and ranged from 2 to 22 pL [11]. Moreover, the patch clamp technique was used to compare the cell condition before and after the injection, and it can be done by applying a series of voltage square pulses from the operational amplifier.

3.2. The Patch Clamp Technique

The principle of the patch clamp technique is to isolate a patch of membrane electrically from the external solution and to record the current flowing into the patch. During each recording, a fire-polished glass pipette filled with a suitable electrolyte solution was pressed against the surface of a cell whilst applying light suction to create a seal whose electrical resistance is more than 1 GΩ, called the gigaseal formation [25]. The patch clamp technique can be carried out under different kinds of configurations, which can record the total current of ion channels on the cell membrane of the intact cell at the whole cell mode [26]. The patch clamp allows an investigation of the change of the cell state from resting potential to action potential and the functional properties of ion channels. The action potential is a transient, regenerative electrical impulse in which the membrane potential (V_m) rapidly rises to a positive voltage value from a negative voltage value, namely the resting potential where the cell is at its resting state [27]. Under a voltage stimulus, the ion channels of the living cell normally respond to generate an electric impulse. The method is believed to be a more precise and accurate way to determine the cell condition.

3.3. SHSY-5Y Cells and HEK-293 Cells

In this study, a SHSY-5Y cell, which is the human derived neuroblastoma cell, and a HEK-293 cell, which is a human embryonic kidney cell, were used to study the performance of the microinjection. The SHSY-5Y cell is often used as in vitro model to study Parkinson's disease through studying the ion channel activities. The HEK-293 cell is widely used in cell biology, including in the study of gene therapy, and is also used as one mammalian expression system in the study of voltage-gated K$^+$ channels [28,29]. The cells were cultured in standard conditions (Dulbecco's Modified Eagle's Medium supplemented with 10% fetal bovine serum and 1% penicillin-streptomycin). The cells were weekly passaged via detachment with trypsin-EDTA, 5-min centrifugation at 1900 rpm, resuspension of the pellet in a 25 mL flask filled with 3 mL of medium, and seeding of a new flask with 100–200 µL of suspension.

3.4. Intracellular and Extracellular Solution

During the micro-injection, the cell lines cultured on a cover slip were bathed in extracellular solution containing 160 mM NaCl, 4.5 mM KCl, 1 mM $MgCl_2$, 2 mM $CaCl_2$, 5 mM of glucose, and 10 mM of HEPES with pH adjustment to pH 7.4 using NaOH. The intracellular solution was prepared using 75 mM KCl, 10 mM NaCl, 70 mM KF, 2 mM $MgCl_2$, 10 mM HEPES, and 10 mM EGTA with pH adjustment to pH 7.2–7.4 using KOH.

4. The Current Response of the Cell Cytoplasmic Microinjection

4.1. The Current Drop of the Cell Cytoplasmic Injection

The cells were placed at the stage of the inverted microscope and optical images were obtained during the injection process, as shown in Figure 5. In each injection, a single cell was selected and a tight seal was formed between the micropipette and the cell surface. Then, the intracellular solution was injected into the cell by injecting around 19 pL intracellular solution.

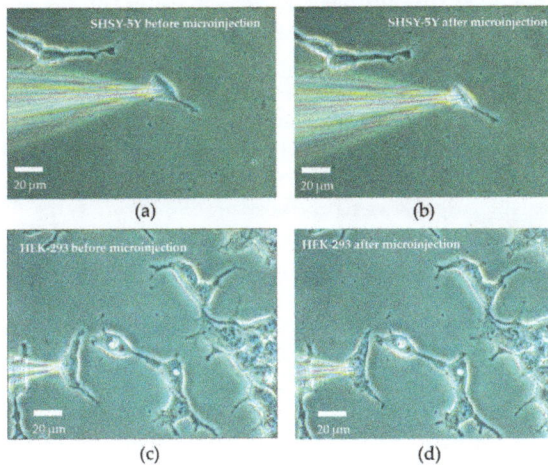

Figure 5. Photos of the micropipette and the cells: (**a**) SHSY-5Y cells before the microinjection; (**b**) SHSY-5Y cells after the microinjection; (**c**) HEK-293 cells before the microinjection; (**d**) HEK-293 cells after the microinjection.

In this work, different batches of SHSY-5Y cells and HEK-293 cells were employed in the experiment. In each batch, one single cell was selected. To elucidate the current response of the SHSY-5Y cells and the HEK-293 cells during the microinjection, current responses of the SHSY-5Y cell and the HEK-293 were recorded as shown in Figure 6. Since the estimated injected volume was 19 μL, the new concentration of [K$^+$] and [Na$^+$] were 150.5 μM and 15.1 μM, respectively, and the $E_{injection}$ was estimated at −0.3 mV, which was not sufficient to activate the ion pumps (<10 mV) after the injection. As a result, $E_{Injection}$ would contribute a current drop during the injection. In this experiment, the accumulated current drop is significant at −11.9 nA and −11.7 nA for the SHSY-5Y cells and HEK-293 cells, respectively.

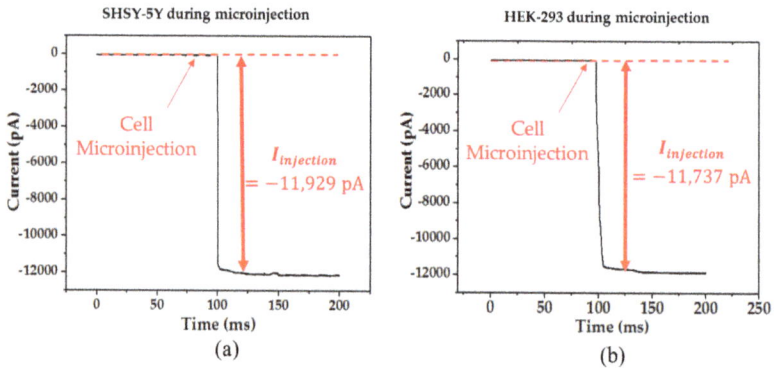

Figure 6. The current response during the cell cytoplasmic microinjection of: (**a**) a SHSY-5Y cell and (**b**) a HEK-293 cell.

4.2. The Current Response for Verifying the Cell Viability after the Cell Cytoplasmic Microinjection

Before and after the microinjection, electrical responses of SHSY-5Y cells were recorded as shown in Figure 7a,b. Similarly, the electrical responses of HEK-293 cells were recorded as shown in Figure 7c,d. The responses of both cells were observed under the voltage stimulus of a square pulse from −80 mV to 100 mV with the epoch of 60 ms. The current responses can be studied through the direction and the amplitude of the current trace.

Before the microinjection, the observation of the normal ion channel activities of both Na$^+$ and K$^+$ ion channels in SHSY-5Y cells were known with the inward current due to the hyperpolarization of Na$^+$ ion channels and the outward current due to the depolarization of K$^+$ channels that occurred at the same time, which was also reported by another research group [27]. Meanwhile, the normal ion channel activities of the K$^+$ ion channels were recorded in HEK-293 before the microinjection.

After the microinjection, the viability can be verified by the current response, due to the ion channel activities evolved by the voltage stimulus of a square pulse from −80 mV to 100 mV with the epoch of 60 ms, applied from the low noise amplifier. The ion channel activities of K$^+$ ion channel were found to be normal with the outward current due to the depolarization of K$^+$ channels [27]. The current responses were compared before and after the microinjection of SHSY-5Y cells and HEK-293 cells. In SHSY-5Y cells, K$^+$ ion channels can be activated in both circumstances, while Na$^+$ ion channels can only be activated before the microinjection. Due to the microinjection, Na$^+$ ions were redistributed across the cell membrane, while Na$^+$ ions inside the cell membrane were consumed and hence Na$^+$ ion channels could not be activated after the microinjection. However, the disappearance of Na$^+$ ions did not lead to the death of the cell, which can be proven by the sustainable activities of the K$^+$ ion channels. It is believed that we can re-activate the Na$^+$ ion channels if we inject more Na$^+$ into the cell. In HEK-293 cells, K$^+$ ion channels can be activated in both circumstances.

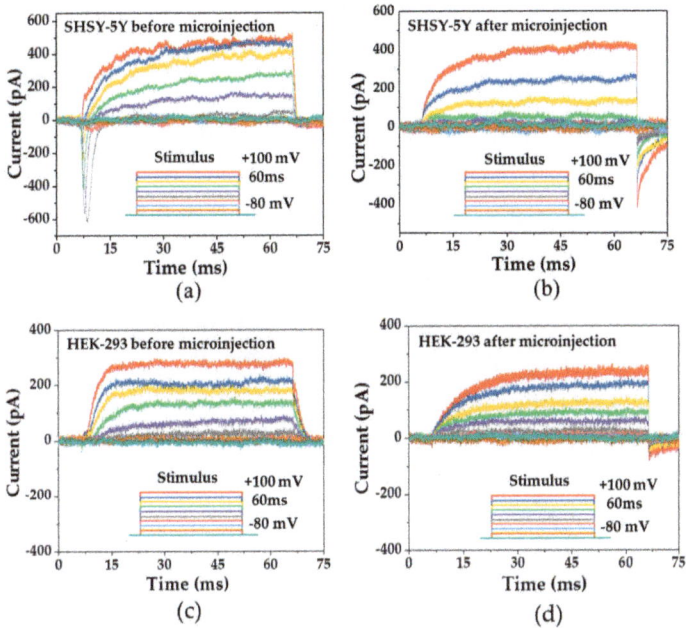

Figure 7. The current traces of ion channels and the voltage stimulus of a SHSY-5Y cell: (**a**) before the microinjection and (**b**) after the microinjection. The current traces of ion channels and the voltage stimulus of a HEK-293 cell: (**c**) before the microinjection and (**d**) after the microinjection.

4.3. Reliability and Repeatability of the Microinjection

In the experiment, the reliability and repeatability are two important indicators. The reliability can be found by the consistency of the total current drop after the injection, whilst the repeatability can be demonstrated by the similar results of two different cells, as shown in Figure 8. We found that the current response of SHSY-5Y cells and HEK-293 cells are (-11.9 ± 0.2) nA and (-11.7 ± 0.4) nA, respectively.

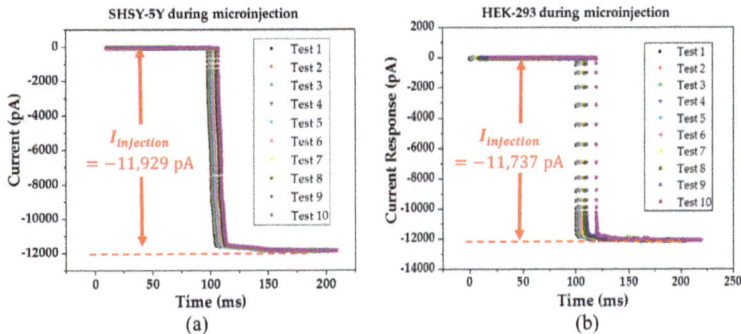

Figure 8. The current response of the cell during the cell cytoplasmic microinjection: (**a**) the microinjection process of the SHSY-5Y cell with 10 results; (**b**) the microinjection process of the HEK-293 cell with 10 results.

Based on the electrical equivalent model, a conceptual control strategy for the microinjection was considered, as shown in Figure 9. Firstly, the position of the cell and pipette were obtained via the visual-based cell and pipette segmentation process, which was developed in our group [30]. Subsequently, the pipette was manipulated to a position on the boundary of the cells, and the automated in vivo whole patch clamping process was then started, which was also developed in our group [31].

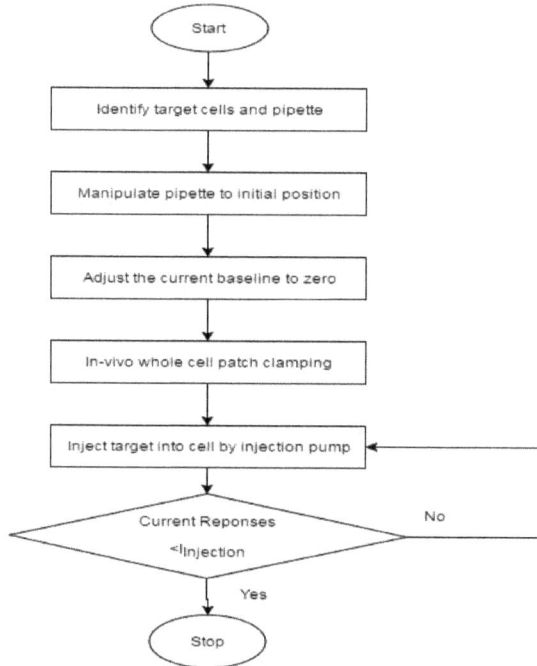

Figure 9. Control strategy of the microinjection system.

Afterwards, the current response during the microinjection was recorded, and the current value was compared with a threshold value. The threshold value is a current drop during the injection ($I_{Injection}$), which we always found during the cell injection. In our experiments, $I_{Injection}$ was found to be at around -11 nA. A current response lower than the threshold can indicate a successful microinjection. In the near future, we will focus on realizing the control strategy in order to build an automated control system.

5. Conclusions

An electric model for a cytoplasmic microinjection with current feedback was designed especially for small adherent cells. To elucidate the model for realizing the current feedback, the cytoplasmic microinjection was performed on the SHSY-5Y cell and the HEK-293 cell. During the microinjection, the consistent current drop (~12 nA) was found on both cells across the cell membrane with a prompt response within several milliseconds for the SHSY-5Y cell and the HEK-293 cell, while the voltage potential of the cell membrane was maintained at -70 mV. The viability of cells after the cytoplasmic microinjection can be well proven by the sustainable activities of K^+ ion channels. In the future, the intracellular solution could act as a transporter of various drugs or other target materials for the micro-injection of small adherent cells using this technique.

Acknowledgments: The research was partially supported by the GRF grant from the Research Grant Council of the Hong Kong Special Administrative Region Government (CityU11205815).

Author Contributions: F.C., R. Y., and K. L. designed the model and experiment. F.C. performed the experiment. F.C. and K.L. analyzed the data and prepared the manuscript.

Conflicts of Interest: The authors declare no conflict of interest.

References

1. Sun, Y.; Nelson, B.J. Biological cell injection using an autonomous microrobotic system. *Int. J. Robot. Res.* **2002**, *21*, 861–868. [CrossRef]
2. Kwon, E.; Kim, Y.; Kim, D. Investigation of penetration force of living cell using an atomic force microscope. *J. Mech. Sci. Technol.* **2009**, *23*, 1932–1938. [CrossRef]
3. Wu, P.H.; Hale, C.M.; Chen, W.C.; Lee, J.S.H.; Tseng, Y.; Wirtz, D. High-throughput ballistic injection nanorheology to measure cell mechanics. *Nat. Protoc.* **2012**, *7*, 155–170. [CrossRef] [PubMed]
4. Gao, J.; Yin, X.F.; Fang, Z.L. Integration of single cell injection, cell lysis, separation and detection of intracellular constituents on a microfluidic chip. *Lab. Chip* **2004**, *4*, 47–52. [CrossRef] [PubMed]
5. Prokop, A.; Davidson, J.M. Nanovehicular intracellular delivery systems. *J. Pharm. Sci.* **2008**, *97*, 3518–3590. [CrossRef] [PubMed]
6. Paulo, C.S.O.; Pires das Neves, R.; Ferreira, L.S. Nanoparticles for intracellular-targeted drug delivery. *Nanotechnology* **2011**, *22*, 494002. [CrossRef] [PubMed]
7. Thomas, C.E.; Ehrhardt, A.; Kay, M.A. Progress and problems with the use of viral vectors for gene therapy. *Nat. Rev. Genet.* **2003**, *4*, 346–358. [CrossRef] [PubMed]
8. Wang, Y.; Yang, Y.; Yan, L.; Kwok, S.Y.; Li, W.; Wang, Z.; Zhu, X.; Zhu, G.; Zhang, W.; Chen, X.; et al. Poking cells for efficient vector-free intracellular delivery. *Nat. Commun.* **2014**, *5*, 4466. [CrossRef] [PubMed]
9. Zhang, Y.; Yu, L.C. Single-cell microinjection technology in cell biology. *BioEssays* **2008**, *30*, 606–610. [CrossRef] [PubMed]
10. Ghanbari, A.; Wang, W.; Hann, C.E.; Chase, J.G.; Chen, X. Cell image recognition and visual servo control for automated cell injection. In Proceedings of the ICARA 2009 4th International Conference on Autonomous Robots and Agents, Wellington, New zealand, 10–12 February 2009; pp. 92–96.
11. Chow, Y.T.; Chen, S.; Wang, R.; Liu, C.; Kong, C.; Li, R.A. Single cell transfection through precise microinjection with quantitatively controlled injection volumes. *Sci. Rep.* **2016**, *6*, 1–9. [CrossRef] [PubMed]
12. Liu, X.; Kim, K.; Zhang, Y.; Sun, Y. Nanonewton force sensing and control in microrobotic cell manipulation. *Int. J. Rob. Res.* **2009**, *28*, 1065–1076. [CrossRef]
13. Huang, H.B.; Sun, D.; Member, S.; Mills, J.K.; Cheng, S.H. Robotic cell injection system with position and force control: Toward automatic batch biomanipulation. *IEEE Trans. Robot.* **2009**, *25*, 727–737. [CrossRef]
14. Karimirad, F.; Chauhan, S.; Shirinzadeh, B. Vision-based force measurement using neural networks for biological cell microinjection. *J. Biomech.* **2014**, *47*, 1157–1163. [CrossRef] [PubMed]
15. Huang, H.; Sun, D.; Mills, J.K.; Li, W.J.; Cheng, S.H. Visual-based impedance control of out-of-plane cell injection systems. *IEEE Trans. Autom. Sci. Eng.* **2009**, *6*, 565–571. [CrossRef]
16. Liu, J.; Siragam, V.; Gong, Z.; Chen, J.; Fridman, M.D.; Ru, C.; Xie, S.; Hamilton, R.M.; Sun, Y. Robotic adherent cell injection for characterizing cell–cell communication. *IEEE Trans. Biomed. Eng.* **2015**, *62*, 119–125. [CrossRef] [PubMed]
17. Permana, S.; Grant, E.; Walker, G.M.; Yoder, J.A. A review of automated microinjection systems for single cells in the embryogenesis stage. *IEEE/ASME Trans. Mechatron.* **2016**, *21*, 2391–2404. [CrossRef]
18. Chow, Y.T.; Chen, S.; Liu, C.; Liu, C.; Li, L.; Kong, C.W.; Cheng, S.H. A high throughput automated microinjection system for human cells with small size. *IEEE/ASME Trans. Mechatron.* **2016**, *21*, 838–850. [CrossRef]
19. Yu, S.; Nelson, B.J. Autonomous injection of biological cells using visual servoing. *Exp. Robot.* **2001**, *7*, 169–178.
20. Xie, Y.; Sun, D.; Liu, C.; Tse, H.Y.; Cheng, S.H. A force control approach to a robot-assisted cell microinjection system. *Int. J. Robot. Res.* **2010**, *29*, 1222–1232. [CrossRef]
21. Seger, R.A.; Actis, P.A.; Catherine, P.; Maaolouf, M.; Vilozny, B.; Pourmand, N. Voltage controlled nano-injection system for single-cell surgery. *Nanoscale* **2012**, *4*, 5843–5846. [CrossRef] [PubMed]

22. Zheng, J.; Trudeau, M.C. *Handbook of Ion Channels*; CRC Press: Boca Raton, FL, USA, 2015; ISBN 9781466551428.
23. Yprey, D.L.; Defelice, L.J. The patch-clamp technique explained and exercised with the use of simple electrical equivalent circuits. In *Electrical Properties of Cells*; Springer: Boston, MA, USA, 2000; p. 7.
24. Hille, B. *Ion Channels of Excitable Membranes*; Sinauer Associates is an imprint of Oxford University Press: Cary, NC, USA, 2001; Volume 507.
25. Martina, M.; Taverna, T. *Patch-Clamp Methods and Protocols*; Humana Press: New York, NY, USA, 2014; ISBN 9781493935642.
26. Sakmann, B.; Neher, E. *Single-channel Recording*; Springer Science & Business Media: New York, NY, USA, 2013; ISBN 9781441912305.
27. Boron, W.F.; Boulpaep, E.L. *Medical Physiology a Cellular and Molecular Approach*; Saunders: Philadelphia, PA, USA, 2013; Volume 53, ISBN 9788578110796.
28. Jiang, B.; Sun, X.; Cao, K.; Wang, R. Endogenous KV channels in human embryonic kidney (HEK-293) cells. *Mol. Cell. Biochem.* **2002**, *238*, 69–79. [CrossRef] [PubMed]
29. He, B.; Soderlund, D.M. Human embryonic kidney (HEK293) cells express endogenous voltage-gated sodium currents and Na v 1.7 sodium channels. *Neurosci. Lett.* **2010**, *469*, 268–272. [CrossRef] [PubMed]
30. Yang, R.; Tam, C.H.; Cheung, K.L.; Wong, K.C.; Xi, N.; Yang, J.; Lai, K.W.C. Cell segmentation and pipette identification for automated patch clamp recording. *Robot. Biomim.* **2014**, *1*, 20. [CrossRef]
31. Yang, R.; Lai, K.W.C.; Xi, N.; Yang, J. Development of automated patch clamp system for electrophysiology. In Proceedings of the 2013 IEEE International Conference on Robotics and Biomimetics, 12–14 December 2013; pp. 2185–2190.

micromachines

MDPI

Article

A Robot-Assisted Cell Manipulation System with an Adaptive Visual Servoing Method

Yu Xie [1], Feng Zeng [1], Wenming Xi [1], Yunlei Zhou [1], Houde Liu [2,*] and Mingliang Chen [3]

[1] School of Aerospace Engineering, Xiamen University, Xiamen 361005, China; xieyu@xmu.edu.cn (Y.X.);
    19920141152946@stu.xmu.edu.cn (F.Z.); wmxi@xmu.edu.cn (W.X.); 19920141152932@stu.xmu.edu.cn (Y.Z.)
[2] Shenzhen Engineering Laboratory of Geometry Measurement Techinology, Graduate School at Shenzhen,
    Tsinghua University, Shenzhen 518055, China
[3] Key Laboratory of Marine Biogenetic Resources, Third Institute of Oceanography, State Oceanic
    Administration, Xiamen 361005, China; mlchen_gg@tio.org.cn
* Correspondence: liu.hd@sz.tsinghua.edu.cn; Tel.: +86-132-4665-8090

Academic Editors: Aaron Ohta and Wenqi Hu
Received: 10 May 2016; Accepted: 15 June 2016; Published: 20 June 2016

Abstract: Robot-assisted cell manipulation is gaining attention for its ability in providing high throughput and high precision cell manipulation for the biological industry. This paper presents a visual servo microrobotic system for cell microinjection. We investigated the automatic cell autofocus method that reduced the complexity of the system. Then, we produced an adaptive visual processing algorithm to detect the location of the cell and micropipette toward the uneven illumination problem. Fourteen microinjection experiments were conducted with zebrafish embryos. A 100% success rate was achieved either in autofocus or embryo detection, which verified the robustness of the proposed automatic cell manipulation system.

Keywords: cell manipulation; robotics; adaptive imaging processing; autofocusing

---

1. Introduction

Microinjecting microliters of genetic material into embryos of model animals is a standard method used for analyzing vertebrate embryonic development and the pathogenic mechanisms of human disease [1,2]. Cell micromanipulation procedure is currently being conducted manually by trained personnel. This requires lengthy training and lack of reproducibility. However, this method cannot meet the demands of the growing development of biological research and the need for testing materials [3]. The integration of robotic technology into biological cell manipulation is an emerging research area that endeavors to improve efficiency, particularly in precision and high throughput aspects.

Recently, several robotic injection prototypes for cell microinjection were reported [4–9]. Wang *et al.* used a position control strategy to inject zebrafish embryos, in which a visual servoing method was used to detect the target position of the end-effector, and a PID (proportional-integral-derivative) position control was used for micropipette movement [4]. Position control with force signal feedback was used by Lu *et al.* to inject zebrafish embryos, where a piezoresistive microforce sensor was used to monitor the injection process [5]. A homemade PVDF (Poly vinylidene fluoride) microforce sensor was proposed in [6] to evaluate the haptic force in a cell injection process. Huang *et al.* used vision and force information to determine three-dimensional cell microinjection, and adopted an impedance control method to control the movement of the injector in the z-direction [7]. Xie *et al.* employed an explicit force control method to regulate the cell injection force on zebrafish embryos [8,9]. However, two problems remained unsolved. First, these studies focused on the motorized injection strategy and control algorithm, even though vision feedback was adopted in every robotic prototype. Past studies did not focus on visual feedback in these robotic injection systems. The segmentation of the embryo and injection pipette is relatively easy for a fixed image. Since invention of automatic microinjection used for large-scale batch microinjections, one of the main challenges lies in the quality of the real-time images that are affected by the environment (*i.e.*, uneven illumination), cell culture medium or individual cell morphology. Therefore, we focused on adaptive and robust image processing in our visual servo system design.

Second, in addition to automating the embryo injection process, a smart visual servoing structure is able to improve the automation level and simplify the whole manipulation system. For instance, a microscope autofocusing system can bring the samples into focus by using the focus algorithm and motion control. To date, no studies concentrated on the autofocusing method for a robot-assisted zebrafish embryo microinjection.

Section 2 of this paper introduces the architecture of the visual servoing cell microinjection robot system. Section 3 reports on the microscope automatic servoing method used to automate the cell manipulation process. The suitability of different focus criteria was evaluated and a visual servoing motion control method is described for a robotic embryo microinjection system. An adaptive visual processing algorithm developed for real-time cell and micropipette location under different illumination environments is discussed. Finally, Sections 4 and 5 report the experimental results and discussion of zebrafish embryos microinjection.

2. The Automatic Microinjection System

2.1. System Configuration

The visual servo cell microinjection system included: (a) a microscope vision processing part; (b) a micromanipulation part; and (c) an integrated interface software platform for visual servoing. The block diagram of the system is shown in Figure 1. The photograph of the microinjection system is shown in Figure 2.

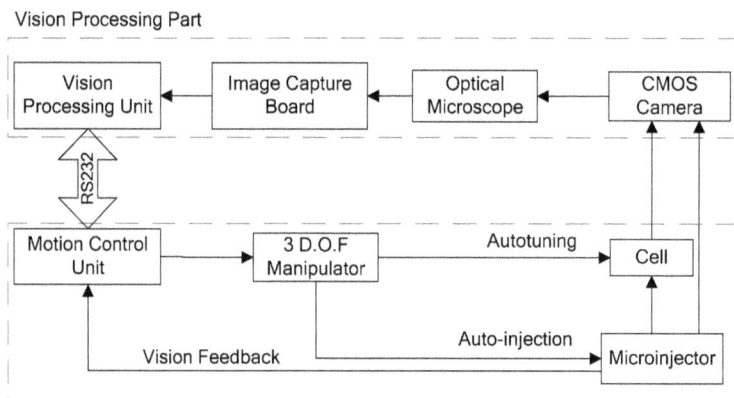

Figure 1. Block diagram of the automatic cell microinjection system.

The visual processing part was responsible for the management of the camera, image acquisition and processing. It included an inverted microscope (model: AE-31, Motic Inc., Wetzlar, Germany) and a CMOS (complementary metal-oxide-semiconductor) camera (model: uEye UI-1540M, IDS Inc., Obersulm, Germany). The microscope had a working distance of 70 mm with a minimum step of 0.2 mm. A CCD (charge-coupled device) adapter of 0.65× and an objective of 4× (N.A. 0.1) were selected to observe the zebrafish embryos. The microscope was working under the bright-field observation mode that provided the necessary optical magnification and illumination levels for proper imaging of the injection area. The CMOS camera was mounted on the microscope with a resolution of 1280 × 1024 pixel, and a 25 fps frame rate was used to acquire the video.

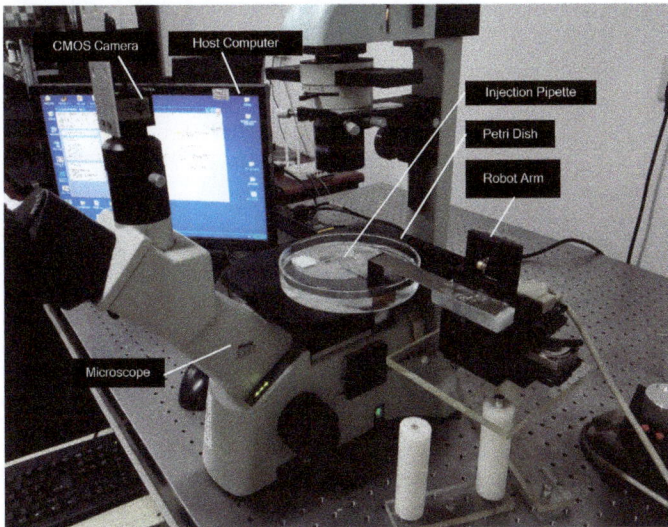

Figure 2. The photograph of the automatic microinjection system.

The micromanipulation part managed the motion controlling instructions from the host computer by handling all processing and signal generation to drive the motion devices using the serial and parallel ports. A three-degrees-of-freedom (3-DOF) robotic arm with a 0.04 μm positioning resolution (model: MP-285, Sutter Inc., Novato, CA, USA) was used to conduct the automatic microinjection task.

To determine the visual servoing automatic microinjection, an integrated software platform was necessary to confirm communications among function modules of the image acquisition, image processing, automatic focusing and automatic microinjection. Because the microinjection system was manipulated from a host computer, a graphical user interface (GUI) was also required to enable the interaction between the user and cell micro-world. More details about the software platform are introduced in the next subsection.

2.2. Integrated Interface Software Platform for Visual Servoing

The integrated interface software platform for visual servoing control was developed under the Microsoft Visual C++ (6.0) environment to ensure the compatibility and portability among the software modulus and hardware. For image acquisition, the camera Software Development Kit (SDK) used C and a small amount of C++ programming, which is compatible with Microsoft Visual C++. For the image processing algorithm, the Intel OpenCV image processing library was used, which also is compatible with Microsoft Visual C++. The 3-DOF manipulator was controlled by a commercial motion controller and was connected to the host computer by an RS-232 serial port. The Windows API (Application Programming Interface) was used to send and receive the position information of the manipulator.

A GUI was designed to provide an interactive way to conduct the robot-assisted microinjection procedure. Figure 3 shows the designed visual servoing microinjection interactive interface developed with the MFC (Microsoft Foundation Classes) framework. The functions of the buttons are described in Table 1.

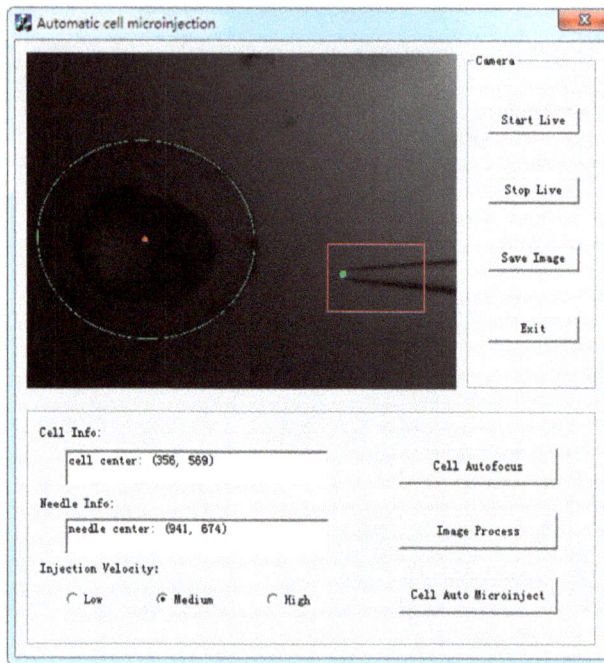

Figure 3. Visual servoing microinjection interactive interface.

Table 1. The button functions of the user interface.

Buttons	Functions
Start Live	Open the camera, and display live video in the picture box
Stop Live	Stop video and display current image in the picture box
Save Image	Save an image to the specified location
Cell Autofocus	Begin automatic cell autofocusing manipulation
Image Process	Begin to search the cell and micropipette by visual processing algorithm and show results in the picture box and corresponding info blocks
Cell Auto Microinject	Begin to automatically move the micropipette and conduct microinjection
Exit	Save and exit the program

3. Visual Servoing Algorithm for Automatic Cell Microinjection

3.1. Visual Servoing Algorithm for Automatic Cell Autofocusing

3.1.1. Selection of the Criterion Function

Since the system was used for a large-scale microinjection, speed and reliability were our primary considerations in the development of the autofocus algorithm because they enhance the efficiency and level of automation for the entire system. Criterion functions were studied for the autofocusing of the microscopes and other optical instruments in prior works [10]. Eighteen focus algorithms were compared in [11,12], where variance based focus algorithms were more sensitive to noise while the gradient-based focus algorithms had better performance in sub-sample cases. In our cell injection system, the image for processing is shown in Figure 3. The zebrafish embryo had a symmetric spherical shape and a clear background, with some dampness from the culture liquid. As such, we narrowed the candidate criterion functions to the following: the Brenner gradient, the Tenenbaum gradient and normalized variance algorithms.

The Brenner gradient [13] measured the differences between a pixel value and its neighbor pixel with a vertical or horizontal distance of two pixel positions. The horizontal direction is used in this paper:

$$f(I) = \sum_x \sum_y \left\{ [I(x+2,y) - I(x,y)]^2 \right\},$$ (1)

where $I(x,y)$ was a gray-level intensity of the pixel at (x,y).

The Tenenbaum gradient (Tenengrad) [14] was a gradient magnitude maximization algorithm that calculated the sum of the squared values of the horizontal and vertical Sobel operators:

$$f(I) = \sum_x \sum_y \left\{ S_x(x,y)^2 + S_y(x,y)^2 \right\},$$ (2)

where $S_x(x,y)$ and $S_y(x,y)$ were the horizontal and vertical Sobel operators.

The normalized variance quantified the differences in the pixel values and the mean pixel value:

$$f(I) = \frac{1}{\mu} \sum_{x,y} (f_{x,y} - \mu)^2,$$ (3)

where μ was the mean pixel value of the image defined in Equation (4).

$$\mu = \frac{1}{N} \sum_x \sum_y I(x,y).$$ (4)

With a selected focus function, the corresponding focus curve was obtained for the captured images along the complete focus interval. Figure 4 shows the normalized focus curves of each image, of which the step length is 200 µm. Different curves arrived at their global peak at the same z-position. All three curves correctly represented the focal plane. Some local maxima were observed with the normalized gradient function. This may prevent the autofocusing algorithm from finding the focal plane or increasing the computational complexity. When compared to the Tenengrade gradient and the normalized variance function, the Brenner function exhibited a more narrow peak, which meant good reproducibility and better searching for the focus plane.

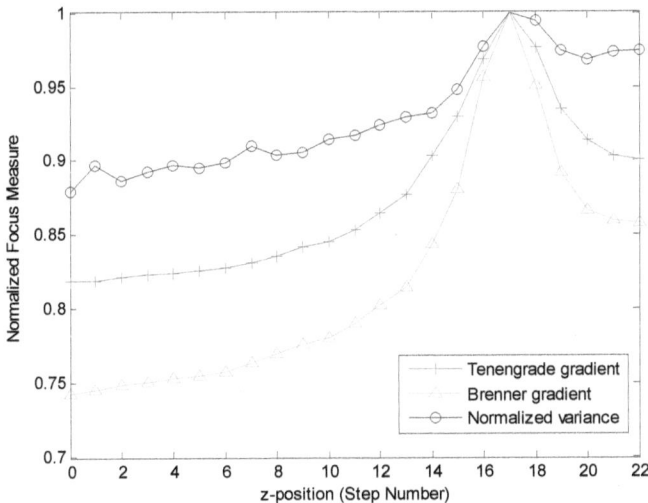

Figure 4. Focus curves after normalization.

Another evaluation criterion for the real-time visual processing system is the computational time of the focus function. A summary of the computational time required to process 23 images is presented in Table 2. The Brenner gradient function took the least time when compared to the other two functions.

Table 2. Computational time for three selected focus functions.

Functions	Tenengrade Gradient Function	Brenner Gradient Function	Normalized Variance Function
Computational Times (s)	1.2794	0.69279	1.0831

Therefore, the Brenner function was chosen as the criteria function for the zebrafish embryos autofocus algorithm.

3.1.2. Implementation of the Automatic Focusing Method

We used an eyepiece with a magnifying power of 10×, the objective of 4×, and a numerical aperture of 0.1. The following equation was used to calculate the depth of field:

$$D_F = \frac{10^{-3}}{7AM} + \frac{\lambda}{2A^2},$$

(5)

where D_F was the depth of field, A was the numerical aperture, M was the total magnification, λ was the light wavelength and the depth of field was $D_F = 63.2$ µm.

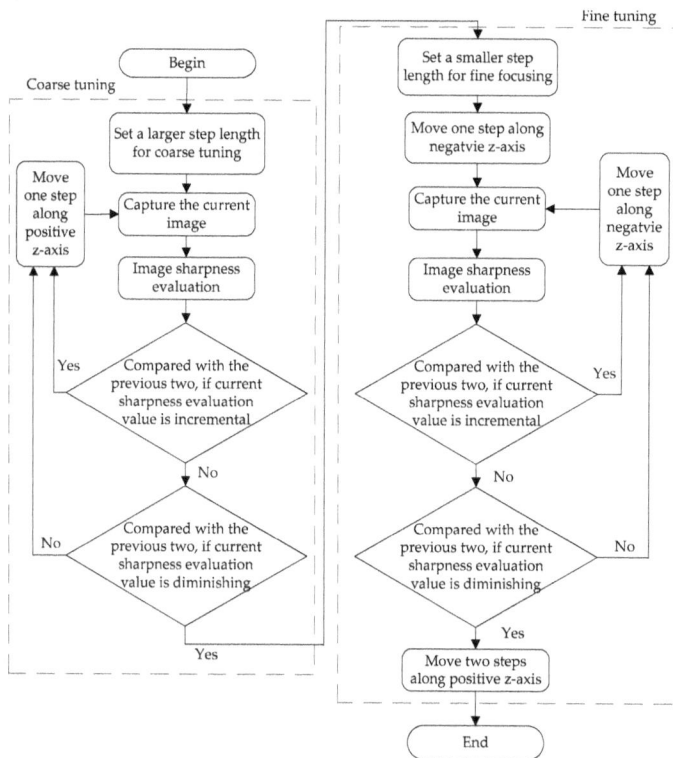

Figure 5. Control method schematic of the entire automatic focusing process.

If the automatic step length was larger than the depth of field, then the cells may move out of the depth of microscope field. Therefore, the step length must be smaller than D_F. To increase the speed of the focusing time, a two-phase automatic focusing method was developed, where a step length of 200 μm was used for coarse focus and 50 μm was used for fine focus. In the coarse focusing phase, the immediate sharpness evaluation value was compared with the previous two images to determine if value was incremental, which would indicate that the manipulator was moving towards the focal plane. If the sharpness evaluation value was not incremental, then it was compared with the previous two images to see if it diminished, which would indicate that the image was out-of-focus. If it was diminished, we began the fine tuning phase, in which the manipulator moved back with a step of fine tuning, similar to the coarse tuning. The control flow of the whole automatic focusing is depicted in Figure 5.

With the Brenner focus function, the corresponding sharpness evaluation value was obtained for the captured images along the complete focus interval. Figure 6a is the coarse focusing curve with a length of 200 μm after normalization. The curve peaked at step 28, which meant it was close to the focal plane. Next, we used a fine focus at step 30 that was also marked as step 0 in the fine focusing stage. The fine focusing curve arrived at a global peak at step 6 that indicated the location of the focal plane, as shown in Figure 6b. The computational time for the Brenner gradient function to process the 31 images in coarse focusing was 0.92786 s, while the time for nine images in fine focusing was 0.23155 s.

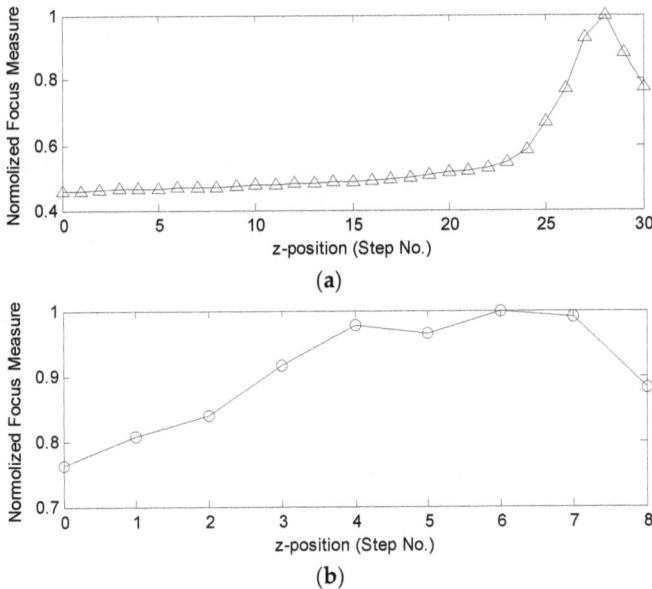

Figure 6. Focusing curves with Brenner gradient function: (**a**) coarse focusing with a step length of 200 μm; and (**b**) fine focusing with a step length of 50 μm.

3.2. Adaptive Image Processing Algorithm for Automatic Cell and Pipette Detection

This section provides real-time location information about the embryo and the injection pipette for the automatic microinjection system. The tasks include (a) detecting and locating the embryo; (b) detecting and locating the injection pipette; and (c) automatically moving the injection pipette to the center of yolk under visual servo. In a real-time automatic cell microinjecting system, one of the primary challenges is the quality of the images affected by the environment (*i.e.*, uneven illumination), cell culture medium or cell morphology. Our algorithm focused on adaptive and robust image processing.

3.2.1. Real Time Adaptive-Threshold Detection for Automatic Cell Detection

The binary operation is a classical threshold segmentation method to separate objects of interest from the background. A conventional binary operation method uses a constant threshold *T* throughout the whole image. Some methods have been proposed to automatically calculate the value, such as the Mean Technique [15], the P-Tile Method [16], the Iterative Threshold Selection [17] and Otsu's method [18]. Figure 7 illustrates the binary operation results using Otsu's method. The conventional threshold was efficient for the uniform illumination images but was not ideal when the illumination was uneven.

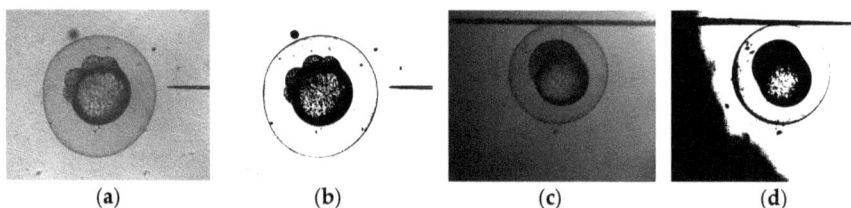

(a)   (b)   (c)   (d)

Figure 7. Results of conventional threshold: even illumination (**left**); uneven illumination (**right**). (a) Original image (even illumination); (b) Binary image with Otsu's method (even illumination); (c) Original image (uneven illumination); (d) Binary image with Otsu's method (uneven illumination).

The images were real-time video images in the automatic injection experiments, so the shadows or the direction of illumination may cause uneven illumination. Non-adaptive methods that analyze the histogram of the entire image are unsuitable. An adaptive threshold calculating method is proposed to specify an adaptive threshold for every pixel in an image. We defined the adaptive threshold as:

$$T_{ij} = A_{ij} - param1, \tag{6}$$

where *A* was the weighted average gray value of pixels in the region around a particular pixel. The block size of the region was represented by parameters of *b* and *param1*.

In this algorithm, pixels with gray value S_{ij} larger than their threshold T_{ij} were set to 255, and all others were set to 0. A circulation for the two parameters was used to adjust the threshold T_{ij} to segment the cell membrane, yolk and background from uneven illumination images. The flow diagram of the circulation to optimize the two parameters is shown in Figure 8. After the adaptive threshold obtained, a regular least squared based ellipse fitting method [19] was used to find the embryo. The contours of the image are detected and every contour is saved in the form of pixel coordinates' vectors of the points. Then, the points in every contour are fitted to the ellipse, which is computed by minimizing the sum of the squares of the distances of the given points to an ellipse. Then, the length of the major axis of the fitted ellipse, *L*, was used to tell if the identified threshold T_{ij} is suitable.

The adaptive-threshold detection method processed different video images in real-time and had good adaptability in both images with uniform illumination and uneven illumination, as shown in Table 3. If the image has more uneven illumination, a bigger block size is required to determine the ellipse (embryo). A green oval was used to mark the embryonic membrane, and a red dot was used to mark the embryo center. The results of our experiments showed that the proposed method can effectively detect edge of the embryo and adaptively locate the embryo center.

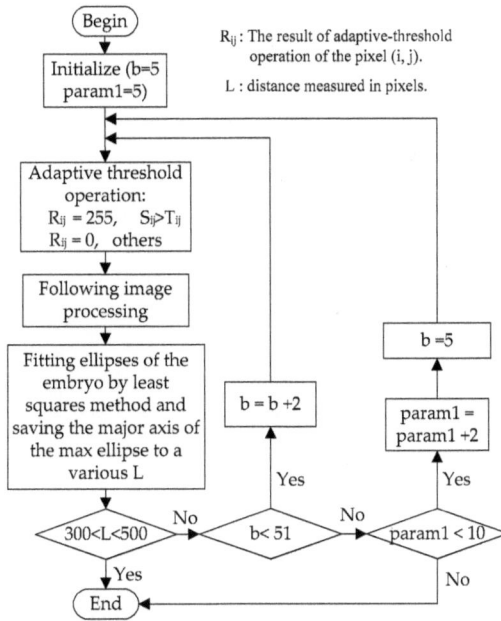

Figure 8. Flow diagram of parameter circulation.

Table 3. Image processing of real-time video images.

Adaptive Threshold	Cell Detection an Location	Center of Embryo	(*b*, *param*1)	Characteristic
		(445, 589)	(7, 7)	even illumination
		(583, 538)	(15, 5)	uneven illumination
		(594, 429)	(13, 5)	uneven illumination, with interference of the image of glass slice
		(549, 556)	(19, 5)	uneven illumination, with interference of the image of another embryo

3.2.2. Detection of Injection Pipette Tip

For a robotic microinjection system, the pipette is fixed on the robot arm. The orientation of the pipette is consequently fixed. The cell microinjection pipette has rigid-bodies with insignificant changes in size and shape along with its movement under the camera. However, the injection pipettes are usually fabricated by a micropipette puller. Even with the same setting parameters as the puller, the size of the pipette may change in a very small scale (*i.e.*, less than 1 μm). We therefore developed an optimized cross-correlation template matching algorithm to track the location of the injection pipette.

First, the tip of the injection pipette was selected as a template $g[k, l]$, and its instance containing the object of interest was detected in a real image $f[i, j]$. We then measured the dissimilarity between the intensity values of the two pictures e as defined by a Sum-of-Squared-Deviations (SSD) template matching:

$$e = \sum_{[i,j] \in R} (f - g)^2. \tag{7}$$

In order to reduce computational cost, Equation (7) was simplified as:

$$\sum_{[i,j] \in R} (f - g)^2 = \sum_{[i,j] \in R} f^2 + \sum_{[i,j] \in R} g^2 - 2 \sum_{[i,j] \in R} f \times g. \tag{8}$$

If f and g are fixed, $\sum f \times g$ measures a mismatch. Therefore, for an $m \times n$ template, we used

$$M[i, j] = \sum_{k=1}^{m} \sum_{l=1}^{n} g[k, l] f[i + k, j + l]. \tag{9}$$

where k and l were the displacements with respect to the templates in the image. For the automatic cell injection, the value f was acquired from the real-time image, which varied from the illumination environment change. To solve this problem, match measure M was optimized as:

$$M[i, j] = \frac{\sum_{k=1}^{m} \sum_{l=1}^{n} g[k, l] f[i + k, j + l]}{\{\sum_{k=1}^{m} \sum_{l=1}^{n} f^2[i + k, j + l]\}^{1/2}}. \tag{10}$$

We calculated the beginning of the image coordinate [1, 1] to find the value of M1, and then calculated the value of the next coordinate [1, 2] as M2. We compare these two values, and recorded the larger value in M1 and recorded the coordinate of the larger value in $T_1[i, j]$. The entire image was searched to find the largest M value. The corresponding coordinate $T[i, j]$ was the location of successful matching.

The coordinate [1, 1] in the template maps to the coordinate $T[i, j]$ in the object image using this template matching algorithm, but the location of the needle tip was still unknown. Therefore, we developed a special template including the relative location of the needle tip, as shown in Figure 9. Then, we could calculate the coordinate of needle tip $[x, y]$, where $x = i + L$, $y = j + H$. Thus, the precise position of the tip in a real-time image was located.

The results of the image processing are shown in Figure 10. A red rectangle was drawn to mark the region of pipette template, and a green dot was used to mark the tip of the injection pipette.

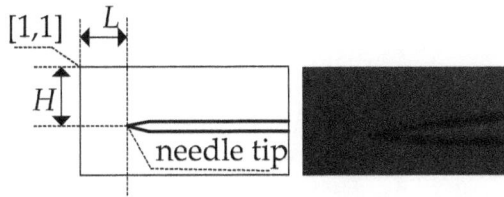

Figure 9. Template structure.

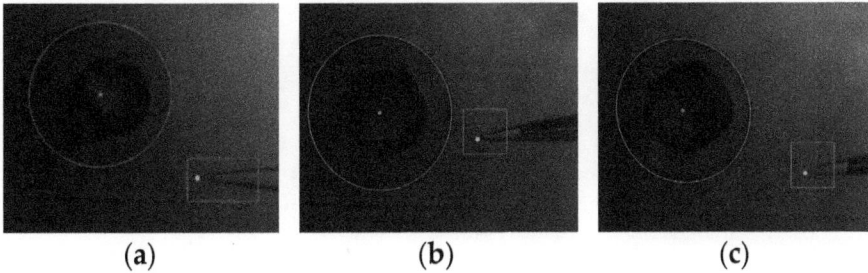

Figure 10. The results of image processing. (**a**) Embryo 1; (**b**) Embryo 2; (**c**) Embryo 3.

4. Experiments

4.1. Materials

The zebrafish embryos were used in the visual servoing cell microinjection experiments, which were grown and collected according to the procedures described in [20]. As shown in Figure 11, the zebrafish embryo was 600–700 µm (without chorion) or 1.15–1.25 mm (with chorion) in diameter, with the cytoplasm and nucleus at the animal pole sitting upon a large mass of yolk. Various chemical substances were released during fertilization, which formed an extracellular space called the perivitelline space (PVS). The injection pipettes were fabricated by a micropipette puller (P2000, Sutter Inc.). The different diameters of the pipette tip were obtained by setting the parameters of laser heating time. Here, the pipettes with tip diameters of 20 µm were selected.

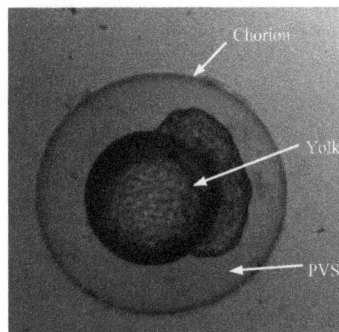

Figure 11. Structure of zebrafish embryo.

4.2. Experiments

Fourteen visual servoed microinjection experiments were conducted to verify the effectiveness of our developed methods. For each embryo, the injection process was as follows:

1. the petri dish containing the embryos and culture medium was placed under the microscope;
2. the embryo was autofocused by using the autofocusing algorithm;
3. the injection pipette was moved to the focus plane;
4. the adaptive image processing was used to get the location and dimension information of the embryo;
5. the template matching algorithm was used to obtain the location of the pipette tip;
6. the distance between the center of the cell and pipette tip along the *x*-axis and *y*-axis was calculated;
7. the injection pipette was automatically moved into center of the embryo;
8. the sample was deposited into the yolk section of the embryo;
9. the pipette out moved of the embryo.

Figure 12 and Video S1 show the typical visual servoing procedures of the microinjection experiments toward the zebrafish embryo. Figure 12a is the image of an embryo after autofocus; Figure 12b shows the successful detection of the embryo and the pipette tip, and Figure 12c is the image of the embryo after automatic injection.

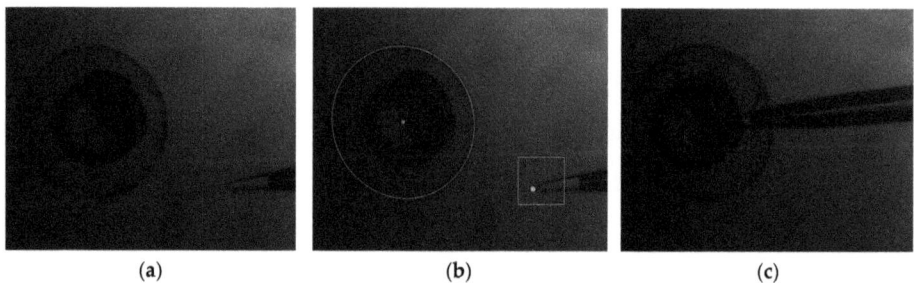

| (a) | (b) | (c) |

Figure 12. Images of automatic microinjection procedure: (**a**) embryo after autofocus; (**b**) detection of the embryo and pipette tip; and (**c**) embryo after injection.

Table 4 shows the results of the automatic microinjection experiments. Every embryo was successfully autofocused using the autofocus algorithm. The major and minor axis lengths show the morphology of the embryo. The second column shows parameters *b* and *param*1 for embryo detection and location. The number of block sizes affects the image processing time since more circulation is needed. With visual processing, the position information was provided for the robotic arm. All fourteen embryos were successfully injected.

Table 4. Results of the automatic microinjection.

Embryo No.	Visual Servoed Autofocus: Success	Embryo and Injection Pipette Detect and Locate			Position Information for Robot Arm		
		The Major and Minor Axis of Embryo (pixel)	Visual Processing with Adaptive-Threshold Algorithm		Target Location of Embryo Center (pixel)	Injection Tip Location (pixel)	Success
			Adaptive-Threshold Parameters (b, $param1$)	Consumption Time (s)			
1	✓	(712, 626)	(37, 5)	2.086	(488, 513)	(969, 247)	✓
2	✓	(726, 636)	(47, 5)	2.824	(326, 442)	(760, 780)	✓
3	✓	(662, 640)	(37, 5)	2.064	(448, 407)	(989, 779)	✓
4	✓	(648, 604)	(45, 5)	2.673	(356, 569)	(917, 677)	✓
5	✓	(672, 646)	(45, 5)	2.651	(407, 590)	(815, 660)	✓
6	✓	(694, 660)	(25, 5)	1.149	(370, 498)	(817, 608)	✓
7	✓	(678, 632)	(21, 5)	0.827	(451, 528)	(884, 483)	✓
8	✓	(648, 614)	(29, 5)	1.482	(393, 481)	(956, 761)	✓
9	✓	(620, 604)	(29, 5)	1.450	(424, 541)	(903, 210)	✓
10	✓	(662, 624)	(35, 5)	1.903	(460, 368)	(1014, 649)	✓
11	✓	(666, 624)	(47, 5)	2.827	(448, 461)	(913, 200)	✓
12	✓	(652, 616)	(35, 5)	1.903	(477, 437)	(893, 280)	✓
13	✓	(648, 602)	(31, 5)	1.619	(391, 459)	(928, 697)	✓
14	✓	(650, 628)	(23, 5)	1.032	(564, 618)	(991, 431)	✓

5. Conclusions

In conventional cell injection, a manual micromanipulation procedure is conducted, which is time-consuming for high throughput and lacks reproducibility. Recently, some semi-automated cell microinjection systems were reported, but they lack robustness and partly rely on human involvement. In our research experiment, we proposed the reduction of human involvement by developing an efficient and adaptive image processing algorithm. Fourteen zebrafish embryos were injected in our experiment, which demonstrated that our system was capable of automatically injecting embryos with a success embryo recognition rate of 100%, and all of the embryos were successfully injected. However, one issue worth noting is that the computational time for the algorithm of adaptively detecting and locating embryos is a bit high. Since the parameters in the adaptive image processing algorithm can be optimized by setting more suitable initial values and developing a more time-saving looping mechanism, the manipulation time can be further reduced.

We designed and used a microrobotic cell manipulation system with an adaptive visual servoing method. We used the Brenner focus algorithm as a criteria function for cell autofocusing manipulation. We also developed an adaptive threshold tuning algorithm for automatic cell microinjection. The cell microinjection system had a 100% success rate using our adaptive imaging processing and microrobotic manipulation control. Future research will develop a knowledge-based automatic cell manipulation method in a complex environment by using deep learning.

Supplementary Materials: The following are available online at http://www.mdpi.com/2072-666X/7/6/104/s1, Video S1: Automatic microinjection.

Acknowledgments: This work was supported by the National Natural Science Foundation of China (No. 61403322) and partly from the Natural Science Foundation of Guangdong Province (2014A030310318).

Author Contributions: X.Y. and H.L. designed the system; F.Z., Y.Z. and M.C. performed the experiments; F.Z. and W.X. designed the algorithms; X.Y. and F.Z. wrote the paper.

Conflicts of Interest: The authors declare no conflict of interest.

References

1. Zon, L.I.; Peterson, R.T. *In vivo* drug discovery in the zebrafish. *Nat. Rev. Drug Discov.* **2005**, *4*, 35–44. [CrossRef] [PubMed]
2. Lin, C.Y.; Chiang, C.Y.; Tsai, H.J. Zebrafish and medaka: New model organisms for modern biomedical research. *J. Biomed. Sci.* **2016**, *23*, 19. [CrossRef] [PubMed]
3. Kari, G.; Rodeck, U.; Dicker, A.P. Zebrafish: An emerging model system for human disease and drug discovery. *Clin. Pharmacol. Ther.* **2007**, *82*, 70–80. [CrossRef] [PubMed]
4. Wang, W.H.; Liu, X.Y.; Gelinas, D.; Ciruna, B.; Sun, Y. A fully automated robotic system for microinjection of zebrafish embryos. *PLoS ONE* **2007**, *2*, e862. [CrossRef] [PubMed]
5. Lu, Z.; Chen, P.C.Y.; Nam, J.; Ge, R.W.; Lin, W. A micromanipulation system with dynamic force-feedback for automatic batch microinjection. *J. Micromech. Microeng.* **2007**, *17*, 314–321. [CrossRef]
6. Pillarisetti, A.; Anjum, W.; Desai, J.P.; Friedman, G.; Brooks, A.D. Force feedback interface for cell injection. In Proceedings of the 1st Joint Eurohaptics Conference and Symposium on Haptic Interfaces for Virtual Environment and Teleoperator Systems—World Haptics Conference, WHC 2005, Pisa, Italy, 18–20 March 2005; pp. 391–400.
7. Huang, H.B.; Sun, D.; Mills, J.K.; Cheng, S.H. Robotic cell injection system with position and force control: Toward automatic batch biomanipulation. *IEEE Trans. Robot.* **2009**, *25*, 727–737. [CrossRef]
8. Xie, Y.; Sun, D.; Liu, C.; Tse, H.Y.; Cheng, S.H. A force control approach to a robot-assisted cell microinjection system. *Int. J. Robot. Res.* **2010**, *29*, 1222–1232. [CrossRef]
9. Xie, Y.; Sun, D.; Tse, H.Y.G.; Liu, C.; Cheng, S.H. Force sensing and manipulation strategy in robot-assisted microinjection on zebrafish embryos. *IEEE/ASME Trans. Mechatron.* **2011**, *16*, 1002–1010. [CrossRef]
10. Groen, F.C.A.; Young, I.T.; Ligthart, G. A comparison of different focus functions for use in autofocus algorithms. *Cytometry* **1985**, *6*, 81–91. [CrossRef] [PubMed]

11. Sun, Y.; Duthaler, S.; Nelson, B.J. Autofocusing in computer microscopy: Selecting the optimal focus algorithm. *Microsc. Res. Tech.* **2004**, *65*, 139–149. [CrossRef] [PubMed]

12. Liu, X.Y.; Wang, W.H.; Sun, Y. Autofocusing for automated microscopic evaluation of blood smear and pap smear. In Proceedings of the 2006 28th Annual International Conference of the IEEE on Engineering in Medicine and Biology Society, New York, NY, USA, 30 August–3 September 2006; pp. 3010–3013.

13. Brenner, J.F.; Dew, B.S.; Horton, J.B.; King, T.; Neurath, P.W.; Selles, W.D. An automated microscope for cytologic research a preliminary evaluation. *J. Histochem. Cytochem.* **1976**, *24*, 100–111. [CrossRef] [PubMed]

14. Krotkov, E. Focusing. *Int. J. Comput. Vision* **1988**, *1*, 223–237. [CrossRef]

15. Al-Amri, S.S.; Kalyankar, N.V.; Khamitkar, S.D. Image segmentation by using threshold techniques. 2010. arXiv:1005.4020.

16. Doyle, W. Operations useful for similarity-invariant pattern recognition. *J. ACM* **1962**, *9*, 259–267. [CrossRef]

17. Jain, R.; Kasturi, R.; Schunck, B.G. *Machine Vision*; McGraw-Hill, Inc.: New York, NY, USA, 1995; pp. 219–6131.

18. Otsu, N. A threshold selection method from gray-level histograms. *Automatica* **1975**, *11*, 23–27.

19. Fitzgibbon, A.; Pilu, M.; Fisher, R.B. Direct least square fitting of ellipses. *IEEE Trans. Pattern Anal. Mach. Intell.* **1999**, *21*, 476–480. [CrossRef]

20. Westerfield, M. A guide for the laboratory use of zebrafish (danio rerio). In *The Zebrafish Book*, 4th ed.; University of Oregon Press: Eugene, OR, USA, 2000.

![micromachines logo] *micromachines*

MDPI

Letter

Fabrication of a Cell Fixation Device for Robotic Cell Microinjection

Yu Xie *, Yunlei Zhou, Wenming Xi, Feng Zeng and Songyue Chen

School of Aerospace Engineering, Xiamen University, Xiamen 361005, China;
19920141152932@stu.xmu.edu.cn (Y.Z.); wmxi@xmu.edu.cn (W.X.);
19920141152946@stu.xmu.edu.cn (F.Z.); s.chen@xmu.edu.cn (S.C.)
* Correspondence: xieyu@xmu.edu.cn; Tel.: +86-151-6006-1679

Academic Editors: Aaron T. Ohta and Wenqi Hu
Received: 21 June 2016; Accepted: 29 July 2016; Published: 4 August 2016

Abstract: Automation of cell microinjection greatly reduces operational difficulty, but cell fixation remains a challenge. Here, we describe an innovative device that solves the fixation problem without single-cell operation. The microarray cylinder is designed with a polydimethylsiloxane (PDMS) material surface to control the contact force between cells and the material. Data show that when the injection velocity exceeds 1.5 mm/s, microinjection success rate is over 80%. The maximum value of the adhesion force between the PDMS plate and the cell is 0.0138 N, and the need can be met in practical use of the robotic microinjection.

Keywords: PDMS; microinjection; robotic; contact force

1. Introduction

Microinjection allows the administration of a foreign target gene directly into the nucleus of a fertilized egg using a glass needle under a microscope to integrate the foreign gene into the recipient cell genome. This is a common biological procedure, used in in vitro fertilization and transgenic technology [1]. Historically, the traditional technique involved transferring cells with a microsuction tube under a microscope and microinjecting with a needle smaller than 10 μm in diameter [2]. However, this required specific skills and time and effort. Recently, Sun and Dong's groups reported successful automation of this technique [3–6]. Specifically, a motor drives the needle to completely inject the cell, and the cell injection force or microscope vision serves as feedback. Robot-assisted microinjection is an emerging area of research that is capable of improving efficiency, particularly in precision and high-throughput aspects.

Current research into automated cell microinjection has been significant, but challenges remain such as cell fixation during injection. Liu et al. made six symmetrical cylinders to fix the cell, and cell force is calculated according to the displacement of the column during the injection process [7]. Lu et al. adopted a gel cell holder with parallel V-grooves and a machine vision algorithm to identify the number of embryos in a batch and then locate the centerline of each embryo [8]. Xie et al. fabricated continuous and transparent V-shape cell structures on piezoelectric sensors, which can be used for cell injection as well as the detection of injection force in the force control system [9]. Liu et al. used microfluid technology to make some cell holding cavities to patch the cells by the differential pressure between the layer above and below. It is a fabulous solution, but the whole system appears to be quite complicated [10]. In Huang et al., suspended cells were fixed by a specially-designed cell holding device that enabled automatic injection of a batch of suspended cells. To facilitate the pick-and-place process of embryos, holes were embedded in circular profile grooves centered around the geometric center of the holder [11]. In sum, the overall fixing methods are based on negative pressure or the fixation structure. However, most of the methods have a common disadvantage—they all need to move the cell to a fixed position. However, it is quite difficult to transfer a single cell into a tiny cell device by manual operation.

To solve the problem, this paper introduced a device based on the adhesive effect of the cell and material surface to make cells fixed without a single cell transfer. Polydimethylsiloxane PDMS has been widely used in the biomedical field as a transparent and nontoxic material [12,13]. Recent research has shown that the contact force between PDMS and an object can be modified by improving the chemical property and surface pattern [14–16]. Wang et al. illustrated the relativity of the friction between the PDMS surface array and a PDMS sphere [17]. This result inspiringly offers us new thought into solving the fixation problem of automatic microinjection. Thus, we fixed a single zebrafish embryo cell using an array. We fashioned microcylinders (1700 × 1700) on a PDMS surface (3.5 cm × 3.5 cm). Cylinders were 10 μm apart. By modifying surface topography, the adhesion force between cells and materials was controlled, and during microinjection, liquid drops containing embryonic cells are placed on the PDMS material using a microsuction tube. In comparison with the traditional fixing methods, a single cell transfer operation is not required, and thus the procedure is simplified to a large extent, which benefits the operator significantly.

2. Experimental Methods

2.1. Surface Texturing

To make the PDMS surface array cylinder, craters (1700 × 1700) are created as a mold on the surface of a silicon wafer. After spin-coating photoresistance onto the wafer and setting the pattern, a wafer mold is etched using an Inductively Coupled Plasma (ICP) deep etching system (AMS200, Alcatel, Annecy, France). The corrosion gas is SF_6 (LinDe Gas, Xiamen, China), the protective gas is C_4H_8 (LinDe Gas). Two kinds of gases act on the silicon wafer, and the etching and passivation are carried out. The etching depth is 5 μm.

PDMS was prepared using a Sylgard 184 (Dow Corning Corp., Midland, MI, USA), and a curing agent was made in a ratio of 12:1 *w/w*. The mixture was placed in a desiccator and degassed for 20 min and then poured into a Petri dish and degassed in a vacuum heating furnace at 70 °C for 12 h. A scalpel was used to evenly cut around the pattern, and the PDMS was removed from the silicon mold. Figure 1 depicts the optical microscope image of PDMS.

Figure 1. Top-view microscope image of polydimethylsiloxane (PDMS).

2.2. Experimental Setup

We tested the fixed performance of the PDMS. Zebrafish embryo cells have different features that are elastic and require different puncture forces for each developmental stage. So, the 32-cell-stage was selected for the microinjection. PDMS was placed in a culture dish submerged in deionized water 5 mm below the water surface. This was to exclude the interference of fluid and surface tension.

A zebrafish embryonic cell mixture was pulled into a microsuction tube and then dripped on the PDMS. After the cell fell on the PDMS surface, the culture dish was then moved to the microscope (AE31, Motic, Xiamen, China) platform to microinject the targeted embryos. To detect the force of the injection, a force sensor (Nano17, ATI, Apex, NC, USA) was applied between it and the motor. A three-degrees-of-freedom (3-DOF) robotic arm with a 0.04 µm positioning resolution (model: MP-285, Sutter Inc., Novato, CA, USA) drives the injection needle to puncture the cell, and cell injection success or failure was recorded. The experimental injection system appears in Figure 2. The whole procedure is shown in Figure 3; the cell was dripped onto the PDMS surface for the microinjection and transferred away by a suction tube.

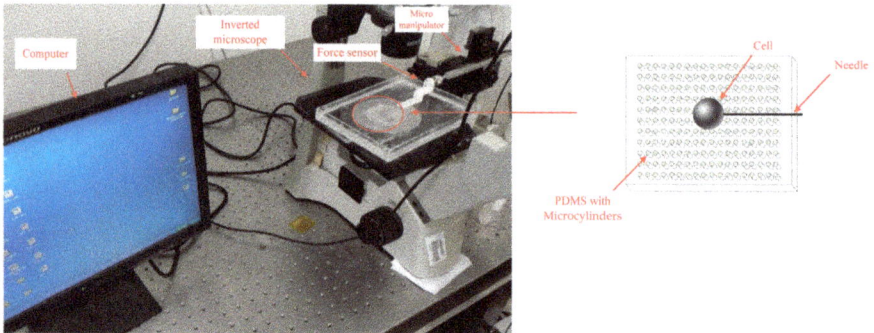

Figure 2. Experimental set-up for zebrafish embryo injection.

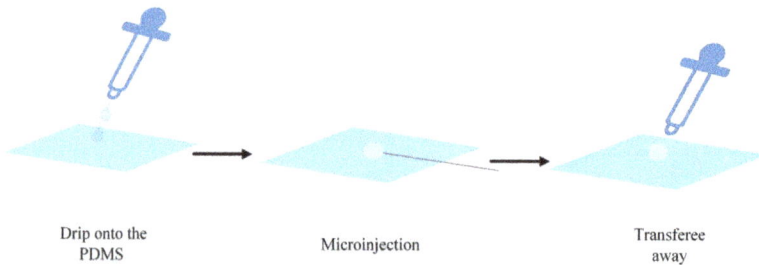

Drip onto the PDMS   Microinjection   Transferee away

Figure 3. The whole procedure process.

The micropipettes are made from a glass tube with an outer diameter of 1.0 mm and an inner diameter of 0.5 mm (TW100F-4, World Precision Instruments, Sarasota, FL, USA), heated and pulled with the use of a micropipette puller (PC-10, Narishige, New York, NY, USA). The micropipette with a tip diameter of 25 µm after grinding is connected to an automatic pressure microinjector (IM-300, Narishige) via a pressure tube.

In the experiment, the cell was injected from the horizontal direction and tested the fixation effect. The angle was chosen because the cell was most prone to move in this condition. Taking the acceleration process into consideration, the injection needle was set 1 mm away from the cell, and the total travel distance during the injection was 1.7 mm.

3. Results

Different injection velocities were tested 20 times. Video S1 shows the robot-assisted microinjection experiment with the zebrafish embryo at 1.5 mm/s. From Figure 4, we can find that the success rate of puncture obviously rises as the injection speed increases. This was the result of the increasing acceleration between the embryo and the needle, as well as the interaction force. Injection success was low when the velocity was less than 1 mm/s, and reached 80% at 1.5 mm/s, which meets the application standard of microinjection. A velocity faster than 1.5 mm/s is feasible, so this method has broad application for automated microinjection processes.

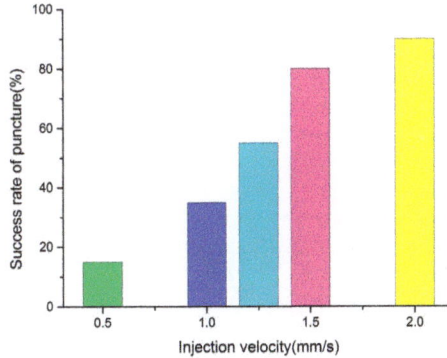

Figure 4. Puncture success rate at different velocities.

A puncture force analysis has been conducted when the velocity is 1.5 mm/s, both in successful and failed cases. Figure 5 shows the case of the contact force between the injection needle and the embryo with a 1.5 mm/s velocity in a failed puncture. The contact force fluctuated when the embryo was exposed to the injection needle and its first movement. This was mainly caused by vibrations of contact and rigid motions. As the injection needle entered the embryo, the embryo membrane was pulled into the embryo, and the contact force between the embryo and the injection needle gradually increased at the same time. When the contact force between the embryo and needle exceeded the contact force between the embryo and the PDMS, the embryo and needle moved together. In this case, the contact force between the embryo and the injection needle was unchanged and both moved together at the same speed.

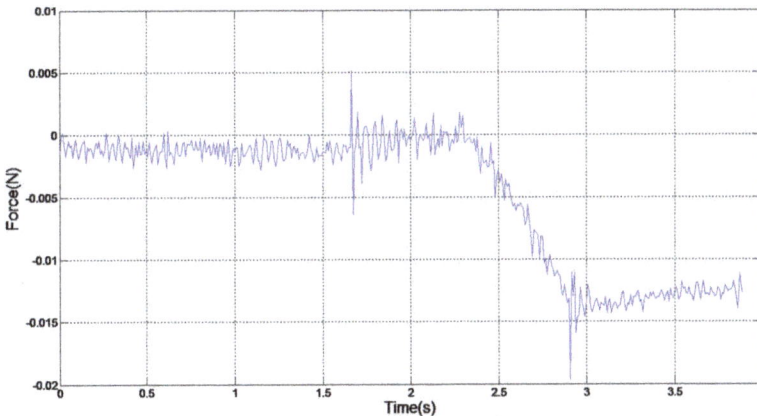

Figure 5. Contact force of failed injection.

Figure 6 shows the classical case of the contact force between the injection needle and the embryo with a 1.5 mm/s velocity in a successful puncture. The contact force gradually increased as the injection needle moved before punctuation and then diminished to indicate successful puncture.

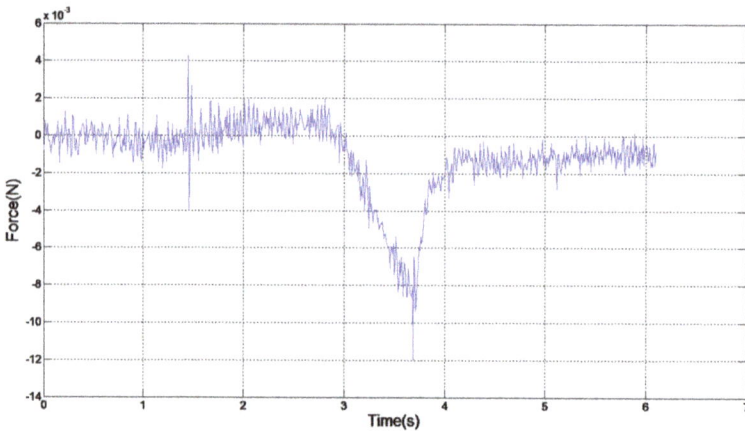

Figure 6. Contact force of successful injection.

From Figures 5 and 6, the key to a successful puncture lies in the embryo puncture force at a certain velocity. In Figure 4, the injection failed, whereas the injection force reached as large as 0.014 N at 1.5 mm/s. However, the embryo was successfully punctured even when the force was 0.009 N at the same velocity. It is normal for a specific zebrafish embryonic embryo to require a large puncture force to penetrate. It is obvious that the adhesion force between the modified PDMS and embryo is limited, and microinjection fails when the injection force exceeds the maximum value of the adhesion force. From the failed group, we calculate 0.0138 N as the maximum value. The diameter of the injection needle also counts in a successful puncture. Less force is needed in the same embryo when the injection needle is thinner. The diameter in usual biomedical experiments is 20 μm, and a size of 25μm is used in this paper; so, the basic standard can be met in practical use. In terms of puncture force, the device we fabricated was again proven to be feasible in robotic microinjection.

4. Conclusions

We fabricated an array cylinder on the light transmitting PDMS material to regulate the force between the material and fixed embryos during automatic zebrafish embryo microinjection. The success rate was 80% when the injection velocity was 1.5 mm/s. The maximum value between the PDMS surface and the cell is 0.0138 N, which meets the requirements in the practical use of robotic microinjection.

Supplementary Materials: The following are available online at http://www.mdpi.com/2072-666X/7/8/131/s1, Video S1: Successful injection at 1.5 mm/s.

Acknowledgments: This work was supported by the National Natural Science Foundation of China (No. 61403322).

Author Contributions: Yunlei Zhou and Xie Yu conceived and designed the experiments; Wenming Xi and Feng Zeng analyzed the data; Songyue Chen contributed materials.

Conflicts of Interest: The authors declare no conflict of interest.

References

1. Morris, A.C.; Eggleston, P.; Cramton, J.M. Genetic transformation of the mosquito Aedesaegypti by micro-injection of DNA. *Med. Vet. Entomol.* **1989**, *3*, 1–7. [CrossRef] [PubMed]
2. Karimirad, F.; Chauhan, S.; Shirinzadeh, B. Vision-based force measurement using neural networks for biological cell microinjection. *J. Biomech.* **2014**, *47*, 1157–1163. [CrossRef] [PubMed]
3. Sun, Y.; Nelson, B.J. Autonomous injection of biological cells using visual servoing. In *Experimental Robotics VII*; Springer: Berlin/Heidelberg, Germany, 2001; pp. 169–178.

4. Sun, Y.; Nelson, B.J. Biological cell injection using an autonomous microrobotic system. *Int. J. Robot. Res.* **2002**, *21*, 861–868. [CrossRef]
5. Wang, W.H.; Liu, X.Y.; Sun, Y. Autonomous zebrafish embryo injection using a microrobotic system. In Proceedings of the 2007 IEEE International Conference on Automation Science and Engineering, Scottsdale, AZ, USA, 22–25 September 2007; pp. 363–368.
6. Huang, H.; Sun, D.; Mills, J.K.; Li, W.J. A visual impedance force control of a robotic cell injection system. In Proceedings of the 2006 IEEE International Conference on Robotics and Biomimetics, Kunming, China, 17–20 December 2006; pp. 233–238.
7. Liu, X.Y.; Kim, K.; Zhang, Y.; Sun, Y. Nanonewton force sensing and control in microrobotic cell manipulation. *Int. J. Robot. Res.* **2009**. [CrossRef]
8. Lu, Z.; Chen, P.C.Y.; Nam, J. A micromanipulation system with dynamic force-feedback for automatic batch microinjection. *J. Micromech. Microeng.* **2007**, *17*, 314. [CrossRef]
9. Xie, Y.; Zhou, Y.L.; Lin, Y.Z.; Wang, L.Y.; Xi, W.M. Development of a Microforce Sensor and Its Array Platform for Robotic Cell Microinjection Force Measurement. *Sensors* **2016**, *16*, 483. [CrossRef] [PubMed]
10. Liu, X.Y.; Sun, Y. Microfabricated glass devices for rapid single cell immobilization in mouse zygote microinjection. *Biomed. Microdevices* **2009**, *11*, 1169–1174. [CrossRef] [PubMed]
11. Huang, H.B.; Sun, D.; Mills, J.K.; Shuk, H.C. Robotic cell injection system with position and force control: Toward automatic batch biomanipulation. *IEEE Trans. Robot.* **2009**, *25*, 727–737. [CrossRef]
12. Chiou, C.H.; Yeh, T.Y.; Lin, J.L. Deformation Analysis of a Pneumatically-Activated Polydimethylsiloxane (PDMS) Membrane and Potential Micro-Pump Applications. *Micromachines* **2015**, *6*, 216–229. [CrossRef]
13. Zhang, L.; Gong, X.; Bao, Y.; Zhao, Y.; Xi, M.; Jiang, C.; Fong, H. Electrospun nanofibrous membranes surface-decorated with silver nanoparticles as flexible and active/sensitive substrates for surface-enhanced Raman scattering. *Langmuir* **2012**, *28*, 14433–14440. [PubMed]
14. Zhou, J.; Khodakov, D.A.; Ellis, A.V. Surface modification for PDMS-based microfluidic devices. *Electrophoresis* **2012**, *33*, 89–104. [CrossRef] [PubMed]
15. Sarvi, F.; Yue, Z.; Hourigan, K. Surface-functionalization of PDMS for potential micro-bioreactor and embryonic stem cell culture applications. *J. Mater. Chem. B* **2013**, *1*, 987–996. [CrossRef]
16. Huang, W.; Wang, X. Biomimetic design of elastomer surface pattern for friction control under wet conditions. *Bioinspir. Biomim.* **2013**, *8*, 046001. [CrossRef] [PubMed]
17. Wang, X.; Wang, J.; Zhang, B. Design principles for the area density of dimple patterns. *Proc. Inst. Mech. Eng. Part J J. Eng. Tribol.* **2015**, *229*, 538–546. [CrossRef]

micromachines

MDPI

Review

A Review on Macroscale and Microscale Cell Lysis Methods

Mohammed Shehadul Islam, Aditya Aryasomayajula and Ponnambalam Ravi Selvaganapathy *

Department of Mechanical Engineering, McMaster University, Hamilton, ON L8S 4L7, Canada;
shehadul.islam@gmail.com (M.S.I.); aryasoma@mcmaster.ca (A.A.)
* Correspondence: selvaga@mcmaster.ca; Tel.: +1-905-525-9140 (ext. 27435)

Academic Editors: Aaron T. Ohta and Wenqi Hu
Received: 21 January 2017; Accepted: 3 March 2017; Published: 8 March 2017

Abstract: The lysis of cells in order to extract the nucleic acids or proteins inside it is a crucial unit operation in biomolecular analysis. This paper presents a critical evaluation of the various methods that are available both in the macro and micro scale for cell lysis. Various types of cells, the structure of their membranes are discussed initially. Then, various methods that are currently used to lyse cells in the macroscale are discussed and compared. Subsequently, popular methods for micro scale cell lysis and different microfluidic devices used are detailed with their advantages and disadvantages. Finally, a comparison of different techniques used in microfluidics platform has been presented which will be helpful to select method for a particular application.

Keywords: cell lysis; cell lysis methods; microfluidics; electrical lysis; mechanical lysis; thermal lysis

1. Introduction

Cell lysis or cellular disruption is a method in which the outer boundary or cell membrane is broken down or destroyed in order to release inter-cellular materials such as DNA, RNA, protein or organelles from a cell. Cell lysis is an important unit operation for molecular diagnostics of pathogens, immunoassays for point of care diagnostics, down streaming processes such as protein purification for studying protein function and structure, cancer diagnostics, drug screening, mRNA transcriptome determination and analysis of the composition of specific proteins, lipids, and nucleic acids individually or as complexes.

Based on the application, cell lysis can be classified as complete or partial. Partial cell lysis is performed in techniques such as patch clamping, which is used for drug testing and studying intracellular ionic currents [1]. In this technique, a glass micropipette is inserted into the cell, rupturing the cell membrane partially. Complete cell lysis is the full disintegration of cell membrane for analyzing DNA, RNA and subcellular components [2].

Different methods have been developed in order to lyse the cell. The nature of lysis method chosen is influenced by the ease of purification steps, the target molecules for analysis, and quality of final products [3]. Laboratory and industrial scale cell lysis methods have been developed and used for many years now. There are a few companies that have also developed equipment (e.g., sonicators and homogenizers) and chemicals (reagents, enzymes and detergents) to lyse cells, which are commercially available. The global market for cell lysis is estimated at 2.35 billion dollars in 2016 and is expected to reach 3.84 billion dollars by 2021 [4].

In past 25 years, conventional laboratory-based, manually-operated bioanalytical processes have been miniaturized and automated by exploiting the advances in microfabrication in the microelectronic industry [5] leading to emergence of a new field known as Microfluidics. Microfluidic technology involves the handling and manipulation of tiny volumes of fluids (nanoliter to picoliter) in the micrometer scale and offers various advantages which include low reagent volume, high surface to volume ratio, low cost and easy handling of small volumes of fluids which are suited for cell analysis. Microfluidic devices have shown great promise in cell lysis and in general cell analysis due to the similar operating size scale [6]. Various researchers have developed microfluidic devices to lyse cells [7]. Researchers have also developed single cell lysis techniques for single cell analysis [8]. This paper reviews several methods of cell lysis techniques that have been used in both macro and micro scale. Finally, a competitive analysis has been performed, which might be helpful to select a process to lyse cell depending on the application and motivation of lysis.

2. Overview of Cell Lysis

Cells are the fundamental unit of all living organisms. Similar to the human body, cells also have a set of organs known as organelles, which are responsible for the cell's ability to perform various kinds of functions. Additionally, the genetic information for the development and functioning of any organism is encoded in DNA or RNA sequences that are located inside the cell. The cell has an outer boundary called cell membrane, which encloses all the contents. The cell membrane serves as a barrier and regulates the transport of material between the inside and outside of the cell. The cell membrane must be disrupted or destroyed in order to access the DNA from inside the cell for molecular diagnosis, such as to identify pathogens [9]. A schematic representation of the cell lysis procedure is shown in Figure 1 where a detergent is used to disrupt the membrane chemically. Detergents react with cell membrane forming pores on the surface of membrane resulting in release of intracellular components such as DNA, RNA, proteins, etc.

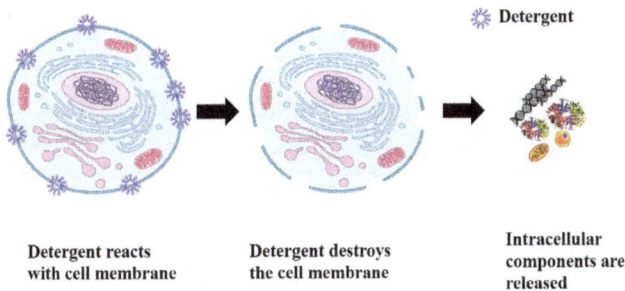

Detergent

Detergent reacts with cell membrane

Detergent destroys the cell membrane

Intracellular components are released

Figure 1. Cell lysis using detergent to open the cell membrane and release the intracellular components. Reproduced with permission from Genomics education program.

2.1. Classification of Cell Types

Cells are of two types: eukaryotic (such as mammalian cells) and prokaryotic (such as bacteria). The main difference between these two types is in their structure and organization.

Figure 2 illustrates the difference between mammalian cells and bacteria. Mammalian cells have a boundary called cytoplasmic membrane that encloses the contents of the cell. In the case of bacteria, there are multiple layers enclosing the cell content and the innermost and outermost of them are called the plasma membrane and cell wall, respectively. Depending on the type of bacteria, the number of these layers varies. In the case of gram-positive bacteria, the plasma membrane is surrounded by another membrane known as cell wall or the peptidoglycan layer, whereas gram-negative bacteria, such as *E. coli*, consist of a cytoplasmic membrane, cell wall and an outer membrane. The composition of these cell layers such as structure and properties, have been extensively reviewed [10–12].

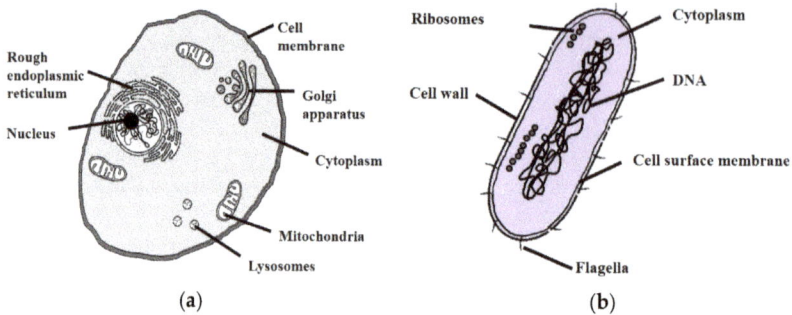

Figure 2. Anatomy of: (**a**) mammalian cell; and (**b**) bacteria.

2.1.1. Cytoplasmic Membrane

Cytoplasmic membrane also known as plasma membrane is a thin structure which acts as a barrier between internal and external environment of cell. This layer is typically 4-nm thick [13,14]. The plasma membrane is mainly made of a phospholipid bilayer that contains highly hydrophobic (fatty acid) and hydrophilic (glycerol) moieties. Figure 3 shows the hydrophobic and hydrophilic configurations of a cell membrane. When these phospholipids aggregate in an aqueous environment, they try to form a bilayer structure where hydrophobic components point to each other and hydrophilic glycerol remain exposed to the outside environment. Proteins are integrated on the surface of the lipid bilayer. Due to the hydrophobic nature of cytoplasmic membrane, it forms a tight barrier; however, some small hydrophobic molecules can pass through this barrier by diffusion.

Figure 3. Structure of cell membrane showing the arrangement of hydrophobic (non-polar) and hydrophilic (polar) regions of phospholipid bilayer.

Eukaryotic cells have rigid and planar molecules called sterols (Figure 4a) in their membrane. The association of sterols increases the stability of cells and makes them inflexible. On the other hand, sterols are not present in prokaryotic cell. However, hopanoids (Figure 4b), molecules similar to sterols, are present in membrane of various bacterial cells. Similar to sterols, hopanoids increase the stability and rigidity of bacterial membrane.

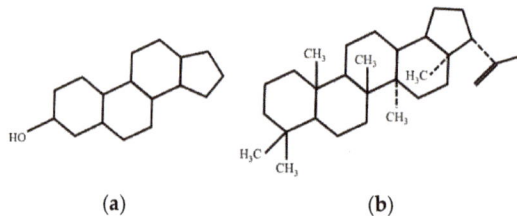

Figure 4. Structure of: (**a**) Sterols; and (**b**) Hopanoids.

2.1.2. Cell Wall

Osmotic pressure is developed inside the cell due to the concentration difference of solutes across the membrane. For *E. coli*, this pressure is estimated around 2 atm [15]. To withstand these pressures, bacteria contains a cell wall or peptidoglycan layer, which also contributes to the shape and rigidity of the cell. This layer consists of two sugar derivatives named *N*-acetylglucosamine and *N*-acetylmuramic acid as well as a small group of amino acids consisting of L-alanine, D-alanine and D-glutamic acid.

The basic structure of this peptidoglycan layer is a thin sheet where the aforementioned sugar derivatives are connected to each other by glycosidic bond forming a glycan chain. These chains are cross-linked by amino acid and the whole structure gives the cell rigidity in all directions. The strength of this structure depends on the frequency of chains and their cross linking.

In gram-positive bacteria, peptidoglycan layer makes up 50%–80% of the cell envelope and 10% of this layer is associated with teichoic acid which provides a greater structural resistance to breakage [16]. In contrast, 10%–20% of the cell envelope of gram-negative bacteria is composed of a 1.2 to 2.0 nm thick peptidoglycan layer [16].

2.1.3. Outer Membrane

In addition to the peptidoglycan layer, there is another layer in the gram-negative bacteria known as the outer membrane. This layer is made of lipopolysaccharide which contains polysaccharides, lipids and proteins. It isolates the peptidoglycan layer from the outer environment and increases the structural firmness of the bacteria. The outer membrane is not permeable to enzymes.

While the focus of the paper is the disruption of the cell boundary, this brief discussion regarding types of cells and their bounding structures is critical in selecting the appropriate methods and materials for lysis. In the next section, the different cell lysis techniques are explained.

3. Classification of Cell Lysis Methods

A number of methods, as depicted in Figure 5, have been established to lyse cells in the macro and micro scale and these methods can be categorized mainly as mechanical and non-mechanical techniques.

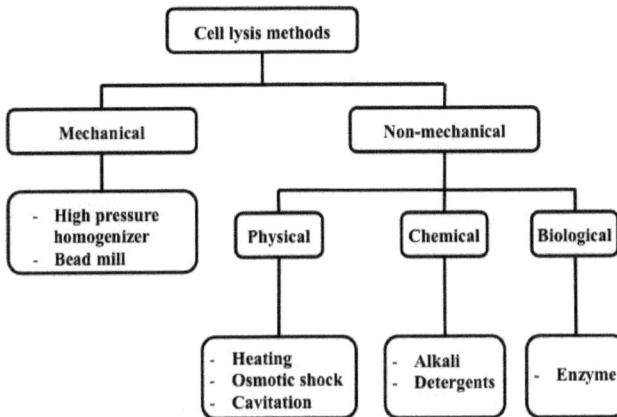

Figure 5. Classification of cell lysis methods.

3.1. Mechanical Lysis

In mechanical lysis, cell membrane is physically broken down by using shear force. This method is the most popular and is available commercially because of a combination of high throughput and higher lysing efficiency. Different types of mechanical lysis techniques are discussed below.

3.1.1. High Pressure Homogenizer

High Pressure Homogenizer (HPH) is one of the most widely used equipment for large scale microbial disruption. In this method, cells in media are forced through an orifice valve using high pressure. Disruption of the membrane occurs due to high shear force at the orifice when the cell is subjected to compression while entering the orifice and expansion upon discharge. Figure 6 shows an example of a commercially available HPH system. In this system, two storage tanks are employed which alternate the feed and allow for multiple passes of the homogenate. A positive displacement pump is used to draw the cell suspension. Depending on the types of the cells, 15–150 MPa is required [3,17].

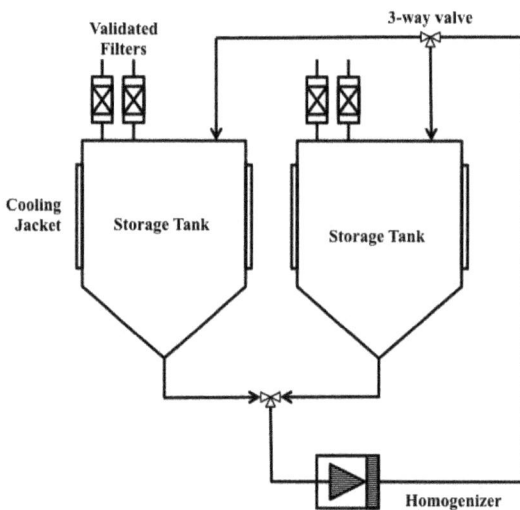

Figure 6. Example of a high pressure homogenizer system. Reproduced with permission from [18].

Sauer et al. [19] proposed a model to relate the amount of protein released by homogenizer to the applied pressure for *E. coli*.

$$\ln\left(\frac{R_m}{R_m - R}\right) = KNP^a \qquad (1)$$

where R is protein released, R_m is the maximum protein available for release, P is pressure in MPa, N is the number of passes, K is the rate constant and a is the pressure exponent.

Since the release of protein is independent of biomass concentration, higher concentration of cell can be disrupted at the same time. However, generation of heat is a problem in this method. Cooling systems can be used to minimize the heat generated. Augenstein et al. [20] reported the degradation of some enzymes during homogenization due to the high pressure. A combination of lysis methods, for example chemical treatment along with homogenization, has shown better results [18].

3.1.2. Bead Mill

Bead mill, also known as bead beating method, is a widely used laboratory scale mechanical cell lysis method. The cells are disrupted by agitating tiny beads made of glass, steel or ceramic which are mixed along with the cell suspension at high speeds. The beads collide with the cells breaking open the cell membrane and releasing the intracellular components by shear force. This process is influenced by many parameters such as bead diameter and density, cell concentration and speed of agitator. Smaller beads with a range of 0.25–0.5 mm are more effective and recommended for lysis [3,21]. Using this technique, several kinds of cells can be lysed for example yeast and bacteria [22,23]. Cell membrane can become totally disintegrated by this method confirming that the intracellular molecules are released. Thus, the efficiency of this method of lysing cells is very high. However, complete disintegration produces small cell debris and thereby separation and purification of sample becomes harder. In addition, heat generation occurs in this process due to the collision between beads and cells. This elevated heat may degrade proteins and RNA.

Ho et al. [24] have compared different cell lysis methods for extracting recombinant hepatitis B core antigen from *E. coli*. They concluded that continuous recycling bead milling method is the most effective method in terms of cost and time. They also report that the most effective method for cell disruption was HPH. Table 1 lists the various commercially available mechanical cell lysis instruments on the market.

Table 1. List of commercially available mechanical cell lysis instruments.

Technique	Trade Names	Website
High pressure homogenizer	DeBEE series Microfluidizer Homogenizers French press G-M	www.beei.com www.microfluidicscorp.com www.gea.com www.glenmills.com
Bead mill	Bead ruptor DYNO MILL TissueLyser II Beadmill Mixer mill Mini bead beater	www.omni-inc.com www.glenmills.com www.qiagen.com www.beadsmill.com www.retsch.com www.biospec.com

Goldberg [25] reviewed the different mechanical cell lysis methods available at both laboratory and industrial scale. Some other mechanical techniques such as rotor/stator shear homogenizer, solid pressure shear, impingement jet and colloid mills are also very efficient in rupturing various kinds of cells [3]. In conclusion, mechanical method is a very efficient method to lyse a wide range of cells. However, problems such as heating of sample volume, degradation of cellular products, cell debris and higher cost limit the use of this method.

3.2. Non-Mechanical Lysis

Non-mechanical lysis can be categorized into three main groups, namely physical, chemical and biological, where each group is further classified based on the specific techniques and methods used for lysis. A detailed description of each type is presented below.

3.2.1. Physical Disruption

Physical disruption is a non-contact method which utilize external force to rupture the cell membrane. The different forces include heat, pressure and sound energy. They can be classified as thermal lysis, cavitation and osmotic shock.

Thermal Lysis

Cell lysis can be conducted by repeated freezing and thawing cycles. This causes formation of ice on the cell membrane which helps in breaking down the cell membrane. This method is time consuming and cannot be used for extracting cellular components sensitive to temperature. Johnson et al. [26] have shown that by using the freeze/thaw cycles they were able to separate highly expressed recombinant proteins from *E. coli*. They submerged the sample solution in dry ice/ethanol bath for 2 min and then thawed in ice/water bath for 8 min. This cycle was repeated three times in total. They compared different cell lysis methods (French press, sonication and enzymatic lysis) and found the freezing/thawing method to be most efficient for extracting these highly expressed proteins. Elevated temperature has also been shown to be capable of cell lysis. High temperature damages the membrane by denaturizing the membrane proteins and results in the release of intracellular organelles. A significant amount of protein can be released from *E. coli* over the temperature range of 90 °C [2,27]. However, heating for a long period may damage the DNA. This method is expensive [28] and so it is not widely used for macroscale industrial applications. In addition, damage of target materials such as protein and enzymes due to higher temperature restricts the use of thermal lysis method. Zhu et al. [29] have described a procedure by modifying the thermal lysis method to extract plasmid DNA from *E. coli* in large quantities (100 mg) in about 2 h. In their method, the *E. coli* are pretreated with lysozyme prior to passing through a heat exchange coil set at 70 °C to lyse the cells. They used peristaltic pump and two heating coils at constant temperature and avoided the use of centrifugation step which enabled them to develop a continuous and controllable flow through protocol for lysing the cells at high throughput and obtaining large quantities of plasmid DNA. Thermal lysis is an attractive method at the micro scale used in many microfluidic devices. The high surface to volume ratio in microfluidic devices helps in cell lysis by quickly dissipating the heat and rupturing the cell membranes effectively. These techniques are covered later in Section 5.

Cavitation

Cavitation is a technique which is used for the formation and subsequent rupture of cavities or bubbles. These cavities can be formed by reducing the local pressure which can be done by increasing the velocity, ultrasonic vibration, etc. Subsequently, reduction of pressure causes the collapse of the cavity or bubble. This pressure fluctuation is of the order of 1000 MPa [3].

During the collapse of a bubble, a large amount of mechanical energy is released in the form of a shockwave that propagates through the media. Since this shock wave has high energy, it has been used to disintegrate the cell membrane. Ultrasonic and hydrodynamic methods have been used for generating cavitation used to disrupt cells.

Ultrasonic Cavitation is a widely known laboratory based technique for disruption of the cells. Ultrasonic vibration (15–20 kHz) can be used to generate a sonic pressure wave [5]. It has been shown that disruption is independent of biomass concentration and proportional to power input. This technique also produces very small cell debris which might be a problem for subsequent processes. In addition, large amount of heat is generated which needs to be dissipated. Enzymes that come out from cell after Ultrasonic Cavitation have also been reported to be degraded [30].

To overcome the problems associated with ultrasonic cavitation, such as high power requirement and high energy to dissipate heat problem, hydrodynamic cavitation has been used to disrupt the cell membrane [31]. Hydrodynamic cavitation is produced by pumping the cell suspension through a constricted channel which results in an increase in velocity. Lee et al. [32] have demonstrated the use of hydrodynamic cavitation as an efficient method to disrupt the cell membrane of cells to extract the lipids. They report that the energy required for lipid extraction from cells using the hydrodynamic cavitation technique was 3 MJ/kg which is 10 times more efficient compared to sonication in terms of energy consumption. In another study by Capocellia et al. [33] the acoustic and hydrodynamic cavitation methods were compared for microbial cell disruption. Their simulation results show that the hydrodynamic cavitation is an order of magnitude more efficient than acoustic cavitation method for cell disruption.

Osmotic Shock

When the concentration of salt surrounding a cell is suddenly changed such that there is a concentration difference between the inside and outside of the cell, the cell membrane becomes permeable to water due to osmosis. If the concentration of salt is lower in the surrounding solution, water enters the cell and the cell swells up and subsequently bursts. This technique is suitable for mammalian cell due to the fragile structure of membrane; however, periplasmic proteins may be released in the case of gram-negative bacteria [34]. Chen et al. [35] compared osmotic shock method and sonication for recovery of recombinant creatinase from *E. coli*. They found osmotic shock method resulted in a 60% creatinase recovery and 3.9 fold purification compared to sonication. They also observed that when the cells were pretreated with divalent cation (Ca^{2+} or Mg^{2+}) the efficiency of osmotic shock method could be improved to 75% and 4.5 fold purification. Another study by Byreddy et al. [36] showed that osmotic shock method resulted in the highest yield of lipids from Thraustochytrid strains when compared to grinding with liquid nitrogen, bead vortexing and sonication methods.

3.2.2. Chemical Cell Disruption

Chemical lysis methods use lysis buffers to disrupt the cell membrane. Lysis buffers break the cell membrane by changing the pH. Detergents can also be added to cell lysis buffers to solubilize the membrane proteins and to rupture the cell membrane to release its contents. Chemical lysis can be classified as alkaline lysis and detergent lysis.

Alkaline Lysis

In alkaline lysis, OH^- ions are the main component used for lysing cell membrane [37]. The lysis buffer consists of sodium hydroxide and sodium dodecyl sulphate (SDS). The OH^- ion reacts with the cell membrane and breaks the fatty acid-glycerol ester bonds and subsequently makes the cell membrane permeable and the SDS solubilizes the proteins and the membrane. The pH range of 11.5–12.5 is preferable for cell lysis [3,38]. Although this method is suitable for all kinds of cells, this process is very slow and takes about 6 to 12 h. This method is mostly used for isolating plasmid DNA from bacteria [39,40].

Detergent Lysis

Detergents also called surfactants have an ability to disrupt the hydrophobic-hydrophilic interactions. Since the cell membrane is a bi-lipid layer made of both hydrophobic and hydrophilic molecules, detergents can be used to disintegrate them. Detergents are capable of disrupting the lipid–lipid, lipid–protein and protein-protein interactions. Based on their charge carrying capacity, they can be divided into cationic, anionic and non-ionic detergents. Detergents are most widely used for lysing mammalian cells. For lysing bacterial cells, first the cell wall has to be broken down in order to access the cell membrane. Detergents are often used along with lysozymes for lysing bacteria (e.g., yeast). Table 2 lists all the detergents according to their charge and properties. Out of the three types of detergents, non-ionic detergents are mostly preferred as they cause the least amount of damage to proteins and enzymes. 3-[(3-cholamidopropyl)dimethylammonio]-1-propanesulfonate (CHAPS) and 3-[(3-cholamidopropyl)dimethylammonio]-2-hydroxy-1-propanesulfonate (CHAPSO), a zwitterionic detergent, is one of the most popular non-ionic detergents. Other non-ionic detergents include Triton-X and Tween series. Ionic detergent such as SDS is widely used for lysing cells because of its high affinity to bind to proteins and denature them quickly. It is used in gel electrophoresis and western blotting techniques. The hydrophilic part of an anionic detergent is mostly a sulphate or carboxylic group whereas for cationic detergent it is ammonium group. Apart from ionic and non-ionic detergents, chaotropic agents can also be used for cell lysis. These include urea, guanidine and Ethylenediaminetetraacetic acid (EDTA) which can break the structure of water and make it less hydrophilic and there by weakening the hydrophobic interactions. An additional purification step has to be in cooperated into the cell lysis protocol when using detergents [41].

Table 2. List of some detergents and their properties. A comprehensive list of detergents can be found here [42].

Detergent	Charge	Properties
Sodium dodecyl sulphate (SDS)	Anionic	Strong lysis agent. Good for most cells. Not suitable for sensitive protein extraction.
Triton X (100, 114)	Non-ionic	Mild lysis agent. Good for protein analysis.
NP-40	Non-ionic	Mild lysis agent. Good for isolating cytoplasmic proteins but not nuclear proteins.
Tween (20, 80)	Non-ionic	Mild lysis agent. Good for cell lysis and protein isolation.
Cetyltrimethylammonium bromide (CTAB)	Cationic	Generally used for isolating plant DNA.
CHAPS, CHAPSO	Zwitterionic	Mild lysis agent. Good for protein isolation.

3.2.3. Enzymatic Cell Lysis

Enzymatic lysis is a biological cell lysis method in which enzymes such as lysozyme, lysostaphin, zymolase, cellulose, protease or glycanase are used. Most of these enzymes are available commercially and can be used for large scale lysis. One advantage of enzymatic lysis is its specificity. For example, lysozymes are used for bacterial cell lysis whereas chitinase can be used for yeast cell lysis and pectinases are used for plant cell lysis. Lysozyme reacts with peptidoglycan layer and breaks the glycosidic bond. For that reason, gram-positive bacteria can be directly exposed to lysozyme, however, outer membrane of the gram-negative bacteria needs to be removed before exposing the peptidoglycan layer to the enzyme. Lysozyme treatment is generally conducted at pH 6–7 and at 35 °C [3]. For gram-negative bacteria, lysozyme is used in combination with detergents to break the cell wall and membrane. Another example is proteinase K which is used for isolating genomic DNA. Andrews et al. [43] and Salazar et al. [44] have reviewed about enzymatic lysis of microbial cells.

3.3. Combination of Mechanical and Non-Mechanical Methods

From the aforementioned discussion, it can be concluded that chemical methods make the membrane permeable which is good for selective product release from cells such as protein or enzymes, however complete cell disruption may not be achieved which may be required for release of other products such as nucleic acid or cell debris. In order to overcome this problem, combinations of non-mechanical and mechanical methods have been employed to increase the efficiency of lysis [3,31]. Anand et al. [45] have studied the effect of combination of mechanical and chemical cell lysis methods on the extent of recovery of intracellular products. They used EDTA as a pretreatment method combined with high pressure homogenizer. The pretreatment lowered the pressure from 34.5 to 13.8 MPa in high pressure homogenizer for recovery of proteins from *E. coli*. cells. They also conclude that pretreatment with guanidium hydrochloride and Triton X-100 resulted in an increase in intracellular release with decrease in usage of energy.

3.4. Overview and Comparison of Different Cell Lysis Methods

A comparison between different types of cell lysis techniques (mechanical and non-mechanical) is summarized in Table 3. It also provides an overview of the major commercial as well as laboratory based lysis techniques with advantages and disadvantages associated with each method.

Table 3. Overview and comparison of cell lysis techniques.

Methods	Equipment and Technique Used	Advantages	Disadvantages
Mechanical	- High pressure homogenizer - Bead mills	- High efficiency - Is not cell dependent	- Heat generated could damage intracellular products - Expensive method - Difficult to purify the lysed sample
Non-mechanical	**Physical**		
	Thermal lysis	- Independent of the cell type - Easy to implement	- Expensive - Damage to proteins and intracellular components
	Cavitation	- Independent of cell type - Large scale integration possible - Operates at a lower temperature and energy level	- Expensive technology - Can cause damage to sensitive proteins - Difficult to purify sample from debris
	Osmotic shock	- Can be used for extracting sensitive intracellular products	- Not suitable for all cell types
	Chemical		
	Alkaline lysis	- Suitable for extraction of sensitive intracellular components (proteins, enzymes, DNA) - Suitable for all kinds of cells	- Slow process (6–12 h)
	Detergent lysis	- Suitable for protein release	- Not suitable for isolating sensitive enzyme and proteins-Expensive reagents - Removal of chemical reagent from sample after lysis is difficult - Lower efficiency as complete lysis is not possible
	Biological		
	Enzymatic lysis	- Can be very specific for cell types - Suitable for extracting proteins	- Complete lysis not possible - Expensive reagents - Has to be used in combination of detergents for bacteria.

4. Microfabricated Platforms for Cell Lysis

Microfluidics is one of the emerging platforms for cell lysis on a micro scale. Microfluidics is the manipulation and handling of small volumes (nano- to picoliters) of liquid in microchannels. Due to the micro scale operation regime, microfluidics is well suited for application where the sample or sample volume is small. This lowers the cost of the analysis due to low consumption of reagents [46]. Microfluidics also enables integration of different modules (or operations) into one device. For example, cells can be lysed and the intracellular products can directly be post processed (PCR or DNA isolation for diagnostics) inside the same device [47,48]. Although there have been a number of reviews on cell lysis in the past 10 years [7,8,49], some of the recent developments in the field have not been reviewed. This review will focus on the recent developments from 2014 onwards and will briefly cover the developments from before, which have been extensively surveyed. Some of the macro scale techniques have been implemented in microfabricated devices for cell lysis. Techniques such as electrical lysis methods are applicable only in the micro scale. Microfluidic lysis technology can be broadly classified into six types. They include mechanical lysis, thermal lysis, chemical lysis, optical lysis, acoustic lysis and electrical lysis.

4.1. Mechanical Lysis

Mechanical lysis in microfluidics involves physically disrupting the cell membrane using shear or frictional forces and compressive stresses. Berasaluce et al. [50] developed a miniaturized bead beating based method to lyse large cell volumes. Zirconium/silica beads were placed inside a cell lysis chamber along with a permanent magnet and actuation of an external magnetic field caused the motion of the beads inside the chamber. Figure 7 shows the various components and device assembled for cell lysis. *Staphylococcus epidermidis* cells were used in this study and they studied the effect of bead size, volume, flow rate and surfactant (Tween-20) on lysing efficiency. They found the optimum parameters achieved a 43% higher yield efficiency at a flow rate of 60 μL/min compared to off chip bead beating system.

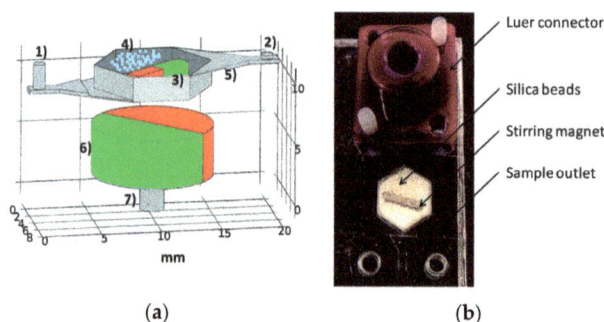

Figure 7. Miniaturized bead beading cell lysis system: (**a**) various components: (1) inlet; (2) outlet; (3) stirring magnet; (4) zirconia/silica beads; (5) bead weir; (6) rotating magnet; and (7) electric motor coupling; and (**b**) image of the device for lysis. Reproduced with permission from [50].

Pham et al. [51] have recently used nanotechnology to fabricate black silicon nano pillars to lyse erythrocytes in about 3 min. They fabricated these nanopillar with ~12 nm tip diameter and 600 nm tall on silicon substrate using reactive ion etching technology. The authors showed that the interaction of erythrocytes cultured on nanopillar arrays causes stress induced cell deformation, rupture and lysis in about 3 min. Figure 8 shows the interaction of erythrocytes with the nanostructures.

Mechanical lysis has been demonstrated by using nano-scale barb [52]. When cells are forced through small opening, high shear forces cause rupture of the cell membrane. Similar principle has been used here where "nanoknives" were fabricated in the wall of microchannels by using modified deep reactive ion etching (DRIE). Distance between these sharp edges was 0.35 μm and width of the channel was 3 μm. The lysis section of this device consisted of an array of these "nanoknives" patterned on a microchannel as shown in Figure 9b. Human promyelocytic leukemia cells (HL-60) were used to pass through this section at sufficient velocity. The addition of this "nanoknives" pattern increased the amount of lysis. This device was used to extract protein from inside the cell. It has been estimated that as much as 99% of the cell was lysed but, only 6% protein was released.

Alternatively, mechanical impingement through collision has also been used to lyse in the microscale [53–55]. Cells were suspended in solution with glass beads and placed on the microfluidic compact disc (CD) device, which was then set to rotate at a very high velocity. The centrifugal force generated by the rotation, causes collision and friction between cells and beads, which results in cell lysis. Various kinds of cells including mammalian, bacteria and yeast have been lysed using this technique.

Though the efficiency of the mechanical lysis is very high, these disruption methods have some drawbacks in microscale application. Fabrication of these devices is complex as well as expensive and collecting the target materials from a complex mixture is very difficult.

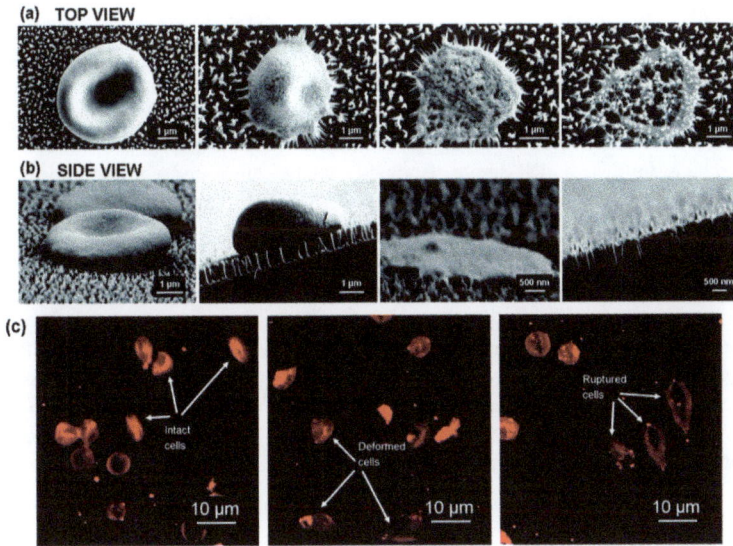

Figure 8. Cell lysis using nano pillars: (**a**,**b**) top and side view of the cells interacting with the nanopillars; and (**c**) confocal laser scanning microscopy pictures of intact, deformed and ruptured cells. Reproduced with permission from [51].

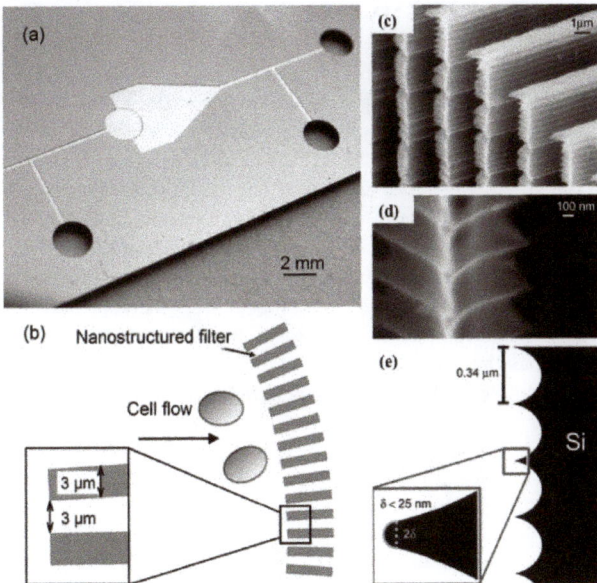

Figure 9. Mechanical lysis using nanoscale barbs: (**a**) microfluidic device showing different inlets and outlet channels; (**b**) schematic of the barbs; (**c**) deep reactive ion etching (DRIE) fabricated nano-knives; (**d**) magnified image of nano-knives patterned using DRIE technique and (**e**) dimensions of the nano-knives used for cell lysis. Reproduced with permission from [52].

4.2. Thermal Lysis

In thermal lysis, heat is supplied to the cells to denature the membrane proteins and lyse the cells. One advantage of thermal lysis is the easy integration of microfluidic devices such as polymerase chain reaction (PCR). The thermal lysis can be performed in such devices with no additional modification. The cells are generally heated above 90 °C and the intracellular products are cycled through different temperatures for example in a PCR device. Tsougeni et al. [56] fabricated a microfluidic device which can capture and lyse cells. They used thermal lysis at 95 °C for 10 min to capture and lyse bacteria. Nanostructures were fabricated in poly(methyl methacrylate) using lithography and plasma etching technique. Microfluidic PCR devices which have incorporated thermal cell lysis [57–59] consist of a glass chamber and a resistive heater to heat the chamber.

In general, thermal lysis is effective in a microfluidic platform, however, these devices are not suitable for sample preparation where the sample is of a large volume and cells have to be lysed from a continuous flow [29]. However, cells have to be treated with lysozyme in order to break the cell wall and make bacteria protoplast. The addition of this lysozyme is time consuming and requires complex structures. Moreover, preserving the enzyme within the device becomes problematic when the device has to be used for a long period of time. Higher lysis time and elevated power consumption are other drawbacks of this method.

4.3. Chemical Lysis

Chemical lysis methods use chemical reagents such as surfactants, lysis buffers and enzymes to solubilize lipids and proteins in the cell membrane to create pores and lyse cells. Although chemical and enzymatic methods are categorized separately in macro scale method, these two techniques are incorporated in the same group for micro scale cell lysis techniques. Buser et al. [60] lysed gram-positive bacteria (Staphylococcus aureus) and RNA virus (respiratory syncytial virus) using a dried enzyme mixture (achromopeptidase). They were able to lyse in less than a minute and then used a disposable chemical heater to deactivate the lysis enzyme. They were able to amplify (off-chip) the lysate without purification and showed the proof of principle for a point of care device for diagnostics.

Kashyap et al. [61] developed a microfluidic probe for selective local lysis of adherent cells (~300 cells) for nucleic acid analysis. Hall et al. [62] used a device for cell lysis experiment, which had two supply wells and a pressure well. Mixing of cell and lysis solution was controlled by adjusting the pressure of the wells. Three different types of solution were used—Solution A containing only SDS (detergent based reagent), Solution B containing surfactant, Triton X-100, Tween-20 with enzyme such as lysozyme, protease, proteinase K and Solution C containing an antibiotic named polymyxin B. Gram-negative and gram-positive bacteria were used for lysis. It was concluded that detergent alone was not suitable for lysis, while Solution B, a mixture of chemical surfactants and biological reagents, can disintegrate the cell membrane and lyse various kinds of bacteria. However, polymyxin B can be potentially used in microfluidic cell lysis platform only for gram-negative bacteria.

Kim et al. [63] also developed a microfluidic device with two inlets and outlets in order to develop an optimal lysis reagent for gram-negative bacteria. Heo et al. [64] demonstrated a microfluidic based bioreactor which was capable of entrapping *E. coli* by using hydrogel patches. Then the immobilized *E. coli* was lysed by using SDS as it can penetrate hydrogel. Cell lysis was accomplished within 20 min. This device was capable of cell lysis using only SDS, however, the previous one could not due to lower exposure time in chemical environment. In another study, Sethu et al. [65] also developed a microfluidic chip (Figure 10) to lyse Erythrocyte in order to isolate Leukocyte. One hundred-percent recovery was possible within 40 s. The device consists of three inlet reservoirs and one outlet reservoir. One inlet was used to flow the entire blood. Second inlet was used for lysis buffer containing mainly aluminum oxide and two side channels were connected with this inlet which converged to direct the entire blood into a narrow stream. This increases the surface contact between the lysis buffer and the cells. The mixture of cells and lysis buffer was then run through a long channel with a number of "U" turns to enhance the buffer. Finally, third inlet was used to flow the phosphate buffer in order to dilute the sample for restoring the physiological concentration [66,67].

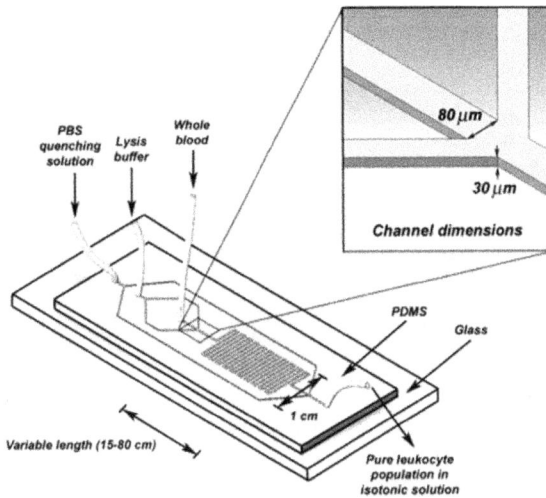

Figure 10. Schematic of a simple chamber and serpentine microfluidic channel for chemical lysis. Reproduced with permission from [65].

Even though chemical lysis method is widely used in many microfluidic devices, this method requires an additional time consuming step for reagents delivery. Therefore, complex microfluidics structures including injection channels and micro-mixers to homogenize the samples are needed [66,68]. After lysis, these reagents might interfere with downstream assay as it is very hard to separate the target molecules [69]. In addition, storage of these reagents is a problem which is why the device cannot be used for long time.

4.4. Optical Lysis

Optical lysis of cells involves the use of lasers and optically induced dielectrophoresis (ODEP) techniques to break open the cell membrane. In laser lysis, a shock wave created by a cavitation bubble, lysis the cell membrane. A focused laser pulse at the cell solution interface creates this cavitation bubble. In ODEP, a conductive electrode and a photoconductive layer (for example amorphous silicon) are formed on the top surface of glass slide. A non-uniform electric field is generated by shining light on the photoconductive layer which then generates a transmembrane potential across the cell membrane disrupting the cell membrane. Huang et al. [70] developed an optically induced cell lysis microfluidic chip for lysing HEK293T cells and extracting intact nucleus. They report cell lysis and nucleus separation efficiency as 78% and 80% respectively using this device.

Kremer et al. [71] lysed cells using an opto-electrical setup. They were able to lyse cells selected based on shape of the cell. They used ODEP to lyse red blood cells in a mixture of red and white blood cells. They developed a method that enabled shape-selectivity such that cells with a different geometry will lyse in a mixture of cell types. The cell with a different shape induces a non-uniform electric field which is used for lysis. Figure 11 shows the schematic of the lysis chip and lysis of differently shaped cells.

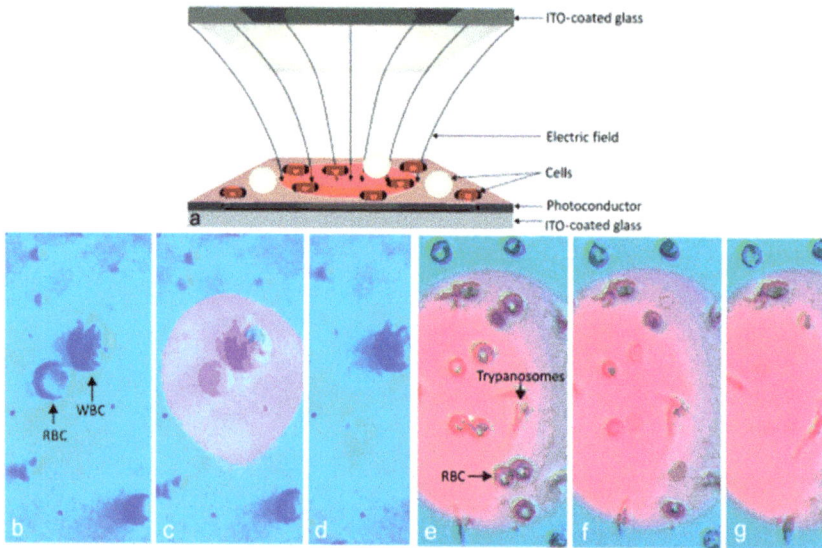

Figure 11. Optical cell lysis device: (**a**) cell lysis chip using optically induced dielectrophoresis (ODEP); (**b**–**d**) cell lysis of red blood cells in a mixture of white and red blood cells; and (**e**–**g**) lysis of red blood cells in a mixture of red blood cells and trypanosomes. Reproduced with permission from [71].

Use of laser light to induce lysis has also been attempted in microfluidic devices. In one instance, optical lysis was induced by application of a nanosecond 532 nm laser pulse [72] which generates a microplasma locally. The plasma collapses causing cavitation, bubble expansion and its collapse as described in previous section are the main reason for a laser induced cell lysis. Various types of cell lines such as rat basophilic leukemia (RBL) [73], rat-kangaroo (Potorous tridactylis) epithelial kidney cells (PtK2) [74], and murine interleukin-3 dependent pro-B (BAF-3) [75] have been lysed by using this laser induced method. However, all these experiments had been done for single cell analysis. It has been found that when laser based lysis was incorporated with polydimethylsiloxane (PDMS) microchannel efficiency of lysis decreased [75]. It was suggested that this may be due to the deformation of PDMS walls which dissipates the mechanical energy from the bubble collapse. For that reason, high energy was required.

Ultraviolet (UV) light array combined with titanium oxide has been used to lyse the cell [76]. Titanium oxide possesses photolytic properties and excitation energy that falls within UV range. When titanium oxides are excited with UV light array, electrons in the valence band are excited to conduct ion band which results in electron–hole pairs. In aqueous environment, these electron–hole pairs react with surrounding molecules and generate free radicals such as OH, O and $O_2{}^-$. These react with cell membrane and lyse the cell. *E. coli* cells were lysed with the above technique. A primary disadvantage of ultraviolet lysis was that the time required to lyse the cell was very high (45 min).

4.5. Acoustic Lysis

In acoustic lysis, a high energy sound wave is generated which is used for cell lysis. This surface acoustic wave (SAW) is produced on a piezoelectric substrate. An inter-digitated transducer (IDT) can be used to produce a SAW electrically with the wave propagating on the surface away from it. Taller et al. [77] have used on chip surface acoustic wave lysis for detecting exosomal RNA for pancreatic cancer study. They achieved a lysis rate of 38% using this technique. Figure 12 shows the fabricated device with the SAW transducer.

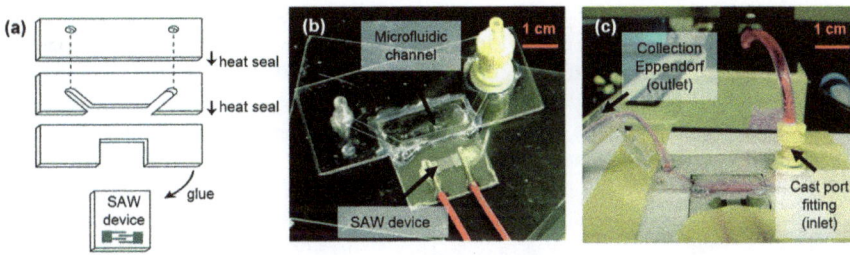

Figure 12. Surface acoustic wave (SAW) lysis microfluidic device: (**a**) assembly of device; and (**b**,**c**) as fabricated device with liquid inlet and outlet for exosome lysis. Reproduced with permission from [77].

They report that the lysis of exosomes is possible due to the effects of acoustic radiation force and dielectric force acting on small particles [78,79]. The SAW device was fabricated using standard photolithography technology. Twenty pairs of titanium aluminum electrodes were patterned on top of piezoelectric lithium niobate substrate to form a single phase unidirectional SAW transducer. This transducer can generate SAW in only one direction. Raw media was exposed to SAW for 30 s at 1 W of power for lysing. The authors report that a lysis efficiency of 38% achieved using this method was sufficient for obtaining enough exosome RNA for detection.

Marentis et al. [80] lysed the eukaryotic cell as well as bacteria by using sonication. This device consists of a microfluidic channel with integrated transducer. The channel was made on glass substrate and piezoelectric transducer was made by depositing zinc-oxide and gold on quartz substrate. The transducers were driven by a sinusoidal source in the 360-MHz range. Eighty-percent lysis of HL-60 and 50% lysis of Bacillus Subtilis spores were obtained by using this device. The temperature rise due to sonication was moderated by using ice pack and cold finger. Ultrasonic horn tip and liquid region are coupled in a microfluidic chip by increasing fluidic pressure in order to increase the efficiency of lysis [81].

Reboud et al. [82] have developed a disposable microfluidic chip to detect the rodent malaria parasite Plasmodium berghei in blood. They used SAW to lyse the red blood cells and parasitic cells in a drop of blood. They report a cell lysis efficiency of more than 99.8% using their device. Xueyong et al. [83] have fabricated a SAW microfluidic device which can lyse red blood cells with high efficiency (95%).

However, sonication has limitations such as generation of heat, complex mechanism as well as expensive fabrication process. Due to this excessive heat generation denaturation of protein and excessive diffusion of the cell contents have been observed [8,84]. To reduce the operation time, cells were first treated with some weak detergent such as digitonin [8,85] before ultrasonic exposure. Digitonin weakened the cell membrane and facilitated lysis.

4.6. Electrical Lysis

In electrical method, cells are lysed by exposing them to a strong electric field. An electric field is applied across the cell membrane which creates a transmembrane potential. A potential higher than the threshold potential is required to form pores in the cell membrane. If the value of the potential is lower than the threshold potential, the pores can be resealed by the cell. On the other hand, a high enough potential can completely disintegrate the cell. At such high voltages, it is found that the electric field does not have any effect on the intracellular components [86]. Electric field is the critical parameter to lyse the cell. As higher electric field is required for cell lysis, high voltage generator is required in order to generate this high electric field in macroscale. Thus, this method is not common in macroscale. However, in microscale due to small size of the devices, higher electric field can be obtained at lower voltage. For this reason and as a method for fast and reagentless procedure of lysis, electrical lysis has achieved substantial popularity in microfluidic community.

Ameri et al. [87] used a direct current (DC) source to lyse cells in a microfluidic chip. Figure 13 shows the fabrication and working principle of their chip. Their device consists of a glass slide coated with indium tin oxide coating patterned for electrodes. The 6400-Microwell arrays are fabricated using SU-8 polymer by photolithography technique. Inlet and outlet channels are created using PDMS polymer and is sealed using a glass slide with ITO electrode for impedance measurement. Red blood cells (10^7 cells/mL) are flown through the device at 20 μL/min and dielectrophoresis (DEP) is used to immobilize the cells into the microarray. A DC voltage of 2 V for 10 s was applied to the cell for lysis. The lysis process was monitored using impedance measurement before and after lysis and a decrease in impedance suggested a complete lysis of cells. They report a lysis efficiency of 87% in their device. The authors proposed a device for cell lysis by electric fields and optical free monitoring of the lysis process on a microfluidic platform which could have potential use in the medical diagnostic field.

Jiang et al. [88] developed a low cost microfluidic device for cell lysis using electric fields. They applied a 10 V square pulse to lyse cells at 50% efficiency. They report a device which had the capability to lyse cells at a much lower voltage compared to a commercially available electropolator device which operated at 1000 V to lyse 200 μL of PK15 cells. They observed bubble formation in their device during cell lysis due to joule heating effect. De Lange et al. [89] have lysed cells in droplets using electric fields. They demonstrated a robust new technique for detergent free cell lysis in droplets. In their device, electric field was applied to lyse bacteria immediately before merging the cell stream with lysozyme and encapsulating the mixture in droplets. They report that with lysozyme alone the lysis efficiency is poor (less than 50%) but when combined with electric fields they were able to obtain up to 90% cell lysis efficiency. Figure 14 shows their microfluidic device for cell lysis in droplets. The authors suggest that their device could be used in applications where use of cell lysis detergents could hinder the cell analysis such as binding assays or studying the chemical activity of proteins and in mass spectroscopy studies where chemical lysis agents can hamper the results.

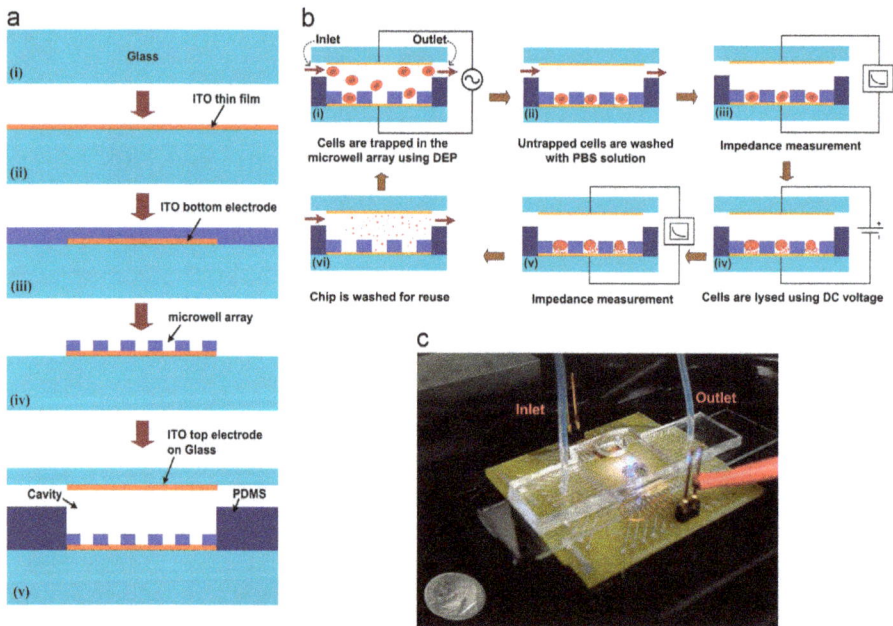

Figure 13. Electrical cell lysis device: (**a**) fabrication protocol of the device; (**b**) working principle of the device; and (**c**) microfluidic device used in the study for lysing red blood cells. Reproduced with permission from [87].

Escobedo et al. [90] showed electrical lysis of cells inside a microfluidic chip using a hand held corona device. They were able to lyse baby hamster kidney cells (BHK), enhanced green fluorescent protein human-CP cells (eGFP HCP) 116 and non-adherent K562 leukemia cells completely inside a microfluidic channel. A metal electrode was embedded inside the channel which was used to discharge 10 to 30 kV to lyse the cells in less than 300 ms. Lysis was assessed by observing before and after images of cells using bright field and high speed microscope and also by cell-viability fluorescence probes. They also report no bubble formation during lysis indicating no joule heating effect thereby making this method suitable for analyzing sensitive proteins and intracellular components. Figure 15 shows the setup and results of the study.

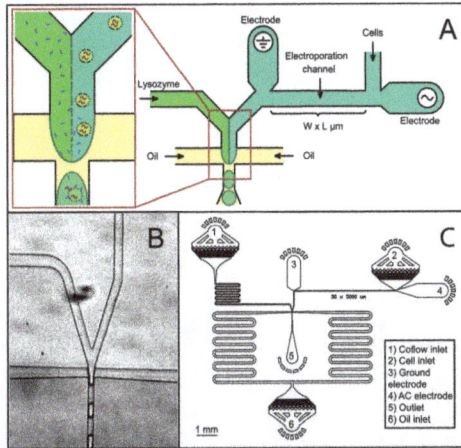

Figure 14. Electrical cell lysis microfluidic device: (**A**) schematic of the electrical lysis and coflow droplet generation microfluidic chip; (**B**) actual image of the droplet generation part; and (**C**) complete electrical lysis with electroporation channels. Reproduced with permission from [89].

Besant et al. [91] detected mRNA molecules of *E. coli* by electrochemical lysis technique. They applied a potential of 20 V, which initiated the cell lysis by producing hydroxide ions from water at cathode to break down bacterial membranes. The sensor electrodes were placed 50 µm away which was enough to detect the mRNA molecules in 10 min. They reported lysis and detection of *E. coli* mRNA at concentrations as low as 0.4 CFU/µL in 2 min which was relevant for clinical application in both sensitivity and time.

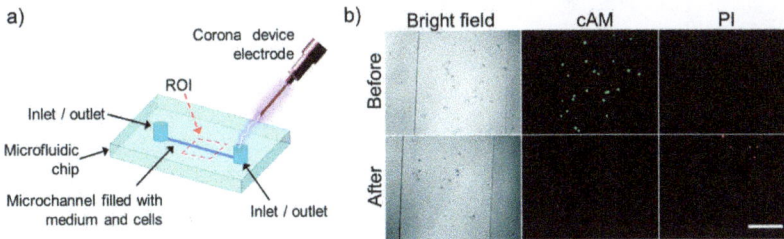

Figure 15. Electrical lysis through handheld plasma device: (**a**) schematic of the device. Cells were lysed using a hand held corona device by applying electric field at the inlet of the device; (**b**) bright field and fluorescent images of before and after of lysis of K562 cells. Reproduced with permission from [90].

Gabardo et al. [92] developed a low cost and easy method to fabricate multi-scale 3D electrodes that could be used for bacterial lysis using a combination of electrical and electrochemical means. These micron-sized electrodes can be rapidly prototyped using craft cutting, polymer induced wrinkling and electro-deposition techniques. They report that these tunable electrodes performed better as compared to lithographically prepared electrodes. They were able to successfully extract nucleic acids extracted from lysed bacteria on a microfluidic platform. They reported 95% lysis efficiency at 4 V using their electrodes. Figure 16 shows the device and electrode structures.

Li et al. [93] developed a double nano-electrode electrical cell lysis device to lyse single neuronal cells. Similarly, Wassermann et al. [94] showed cell specific lysis of up to 75% of the total human blood cells using SiO_2 passivated electrical cell lysis electrodes at an applied voltage of 8–20 V. Ma et al. [95] reported a 10–20-fold increase in mRNA extracted from *M. smegmatis* using electrical lysis in a microfluidic platform as compared to a commercial bead beading instrument. They used a 4000–8000 V/cm field intensity to lyse the bacteria with long pulses (5 s). They report that their device can be effective for mRNA release from hard to lyse cells.

Islam et al. [96] showed the proof of concept of a simple microfluidic device for electrical lysis of larger volumes of sample. They used a nanoporous membrane sandwiched between two microfluidic channels to trap and lyse *E. coli* bacteria by applying 300 V. They report a lysis efficiency of 90% in less than 3 min. Figure 17 shows the schematic of the device used for lysis in their study.

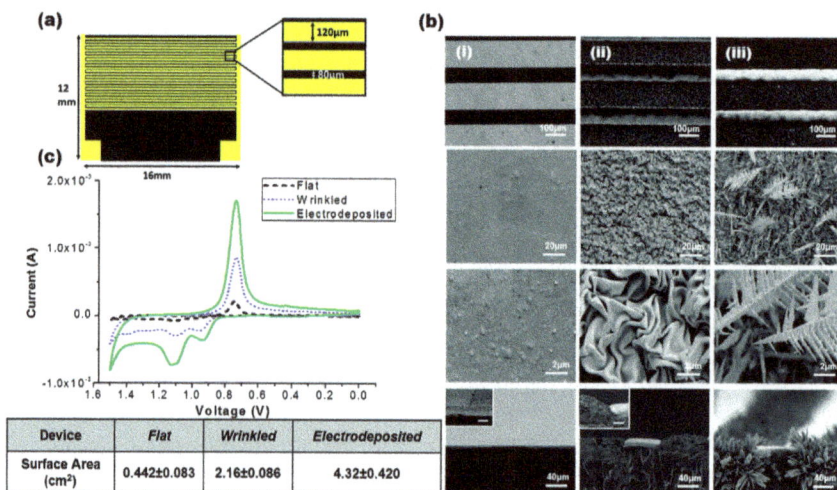

Figure 16. Bacterial lysis device: (**a**) schematic of the lysis device; (**b**) scanning electron micrographs of: (**i**) planar; (**ii**) wrinkled; and (**iii**) electrodeposited electrodes; (**c**) cyclic voltammetry scan of the electrodes. Reproduced with permission from [92].

Different types of voltages such as alternating current (AC) [97,98], DC pulses [99–101] and continuous DC voltages [102] have been used in order to lyse the cells. Along with electric field, exposure time of cells within that electric field is also an important parameter for cell lysis. It has been found that cells can be lysed by using higher electric field for short period of time as well as lower electric field for long period of time [103]. For that reason, AC and DC pulses of a higher electric field are needed as compared to a continuous DC electric field. As the electric field depends on the distance between the electrodes, microfabricated electrodes have been used during AC or DC pulses. An overview of different electrical lysis devices and the characteristics of the designed system is presented in Table 4.

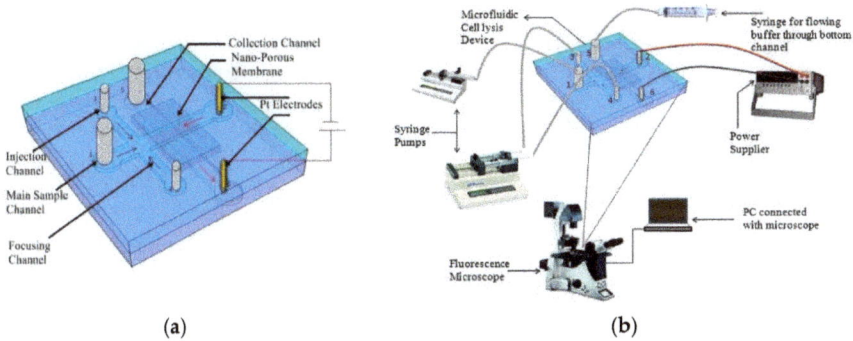

Figure 17. Electrical cell lysis microfluidic device: (**a**) schematic of cell lysis device; and (**b**) experimental setup. Reproduced with permission from [96].

Table 4. Different electrical lysis devices used for cell lysis.

Reference	Species	Type of Cell	Cell Size (μm)	Electrode	Type of Voltage (AC/DC)	Lysing Voltage (V)
[104]	Human	HT-29	10	Gold	AC	8.5
[97]	Human	A431	10	Gold	AC	20
[98]	-	FITC-BSA laden vesicle	50	ITO	AC	5
[99]	-	Leukocytes	-	3D	DC pulse	10
[100]	Human	Red blood cells	6–8	Pt wire	DC/AC	30–170
[105]	Bacteria	E. coli	-	Gold	DC pulse	50
[106]	Hamster	CHO	10–16	Pt wire	DC pulse	1200
[106]	Bacteria	E. coli		Pt wire	DC	930
[102]	Human	Red blood cells	6–8	Pt wire	DC	50
[87]	Human	Red blood cells	6–8	ITO	DC	2
[96]	Bacteria	E. coli	-	Pt wire	DC	300

Lu et al. [104] developed a microfluidic electroporation platform in order to lyse human HT-29 cell. Microfabricated saw-tooth electrode array was used in order to intensify the electric field periodically along the channel. Seventy-four-percent efficiency was obtained for an operational voltage of 8.5 V. However, this mode of lysis is not suitable for bacteria due their sizes and shapes. Compared to mammalian cell, high electric field and longer exposure is needed to lyse bacteria. Rosa [105] developed a chip to lyse bacteria consisting of an array of circular gold electrodes. DC pulses were used and lysis with 17% efficiency was achieved by using an operational voltage 300 V. This efficiency was increased up to 80% after adding enzyme with cell solution. In 2006, Wang et al. [107] proposed application of continuous DC voltage along the channel for cell lysis. The device consists of a single channel with uniform depth and variable width. Since the electric field is inversely proportional to width of the channel, high electric field can be obtained at the narrow section of the channel. Thus, lysis occurs into a predetermined portion of the device. Exposure time of the cell to the electric field can be tuned by changing the length of this narrow section. The configuration of the device was optimized and lysis of complete E. coli bacteria was possible at 930 V. Complete disintegration of cell membrane was observed when the electric field was higher than 1500 V/cm. This device was very simple and did not need any microfabricated electrodes. Pt wires were used as electrodes. Only a power generator was needed to operate it. However, bubble generation and Joule heating issue could not be completely eliminated. Similar kind of device was used by Lee [102] where the length and width of the narrow section was modified in order to lyse mammalian cell. Bao et al. [108] also developed a device to lyse E. coli by using DC pulses. Release of intracellular materials was observed when the electric field was higher than 1000 V/cm.

In conclusion, electrical method offers a simple, fast and reagent less lysis procedure to lyse various kinds of cells. This method is also suitable for selective lysis and is compatible with other downstream assays such as amplification and separation. Although requirement of high voltage is a problem in this procedure, it can be overcome by decreasing the gap between electrodes through microfabrication. However, heat generation and formation of bubble is a major problem for electric lysis method.

4.7. Comparison of Different Microfluidic Technologies for Cell Lysis

Various microfluidic technologies for cell lysis are compared in Table 5. The advantages and disadvantages of different methods are listed for each technique.

Table 5. Comparison of different microfluidic lysis methods. Cell lysis efficiency was determined by averaging the lysis efficiencies from the references cited. Low: 0%–50%; Medium: 50%–80%; High: 80%–100%.

Lysis Method	Lysis Time	Efficiency	Pros and Cons
Mechanical	30 s–10 min	Medium	- Can lyse any type of cell - Robust and can be used for tough cells - Complex device fabrication - Expensive method
Thermal	2–5 min	High	- Easy to integrate into device - Low cell lysis time-High power consumption - Cannot be used for extracting sensitive intracellular components.
Chemical	30 s–20 min	High	- Low cost and lysis time - Additional step of adding the chemical - Expensive lysis reagents
Optical	30 s–10 min	High	- Single cell lysis is possible - Cell specific lysis - Expensive and complex instrumentation required - Slow lysis time
Acoustic	3 s–1 min	Medium	- Easy integration of electrodes into microfluidic device - Fast cell lysis time - Expensive technology - Heat generation during lysis
Electrical	50 ms–10 min	High	- Fast lysis time - Easy integration of electrodes in device - Joule heating - Expensive equipment for lysis

5. Single Cell Lysis

Single cell analysis has gained much popularity in the recent years owing to the development of new technology. Single cell analysis can be used to understand the cellular heterogeneity in a cell culture as well as used in popular areas of genomics, transcriptomics, proteomics and metabolomics. Single lysis is one of the first steps involved in single cell analysis of intracellular components (proteins, enzymes, DNA, etc.). Many different platforms have been used to study single cell lysis including microfluidics, high speed imaging, capillary electrophoresis and PCR. Cell lysis methods such as laser pulse, nanoscale barbs, acoustic, electrical and chemical (detergents and enzymes) have been utilized to lyse cells. Brown et al. [8] have reviewed single cell lysis methods extensively.

Single cell lysis buffers offered commercially are optimized for single cell RNA extraction. These buffers are designed to reduce sample loss and are compatible with enzymatic reactions such as reverse transcription. Single cell lysis buffers are commercially available from companies such as Thermo Fisher Scientific Inc. (Runcorn, UK), Takara Bio Company (Otsu, Japan), Bio-Rad Laboratories Inc. (Hercules, CA, USA), Signosis (Santa Clara, CA, USA), etc. Companies such as Fluidigm, Dolomite Bio and Molecular Machines and Industries (MMI) specialize in single cell lysis equipment for genomics studies. Svec et al. [109] compared 17 different direct cell lysis protocols for transcript yield and compatibility using quantitative real time PCR method. They concluded that bovine serum albumin (BSA) resulted in the best lysis reagent which resulted in the maximum lysis efficiency and high RNA stability. Kemmerling et al. [110] designed a microcapillary electrode to lyse single cells using electrical pulses. The cell lysates were aspirated into the microcapillary to be later analyzed directly in a transmission electron microscope for protein analysis. Developments in single cell analysis technologies have opened up new possibilities and discoveries in the area of genomics and proteomics.

6. Summary

This review provides an overview of cell lysis techniques in the macro and micro scale. The macroscale cell lysis techniques are well established and commercialized by many companies. These techniques include mechanical, chemical, physical and biological techniques. On the other hand, microscale and single cell lysis techniques have recently evolved and use the same macroscale principles for lysis in a miniaturized device. The choice of cell lysis method depends on the type of cells, concentration, application (post processing) and efficiency required. It is difficult to choose one technology, since each method has its own advantages and disadvantages. This review provides a guideline for researchers to choose the cell lysis technology specific for their application. As novel fabrication techniques are introduced in the microfluidics field, we will see better cell lysis techniques with higher efficiency and faster lysis times at reduced cost.

Acknowledgments: The authors acknowledge funding support from the Ontario Research Fund - Research Excellence Program, the Natural Sciences and Engineering Research Council of Canada through their Discovery Program, and the Canada Research Chairs Program.

Conflicts of Interest: The authors declare no conflict of interest.

Abbreviations

PCR	Polymerase Chain Reaction
E. coli	*Escherichia coli*

References

1. Sakmann, B.; Neher, E. Patch clamp techniques for studying ionic channels in excitable membranes. *Annu. Rev. Physiol.* **1984**, *46*, 455–472. [CrossRef] [PubMed]
2. Goodfellow, M.; Stackebrandt, E. *Nucleic Acid Techniques in Bacterial Systematics*; John Wiley & Sons: Hoboken, NJ, USA, 1991; Volume 5.
3. Harrison, S.T.L. Bacterial cell disruption: A key unit operation in the recovery of intracellular products. *Biotechnol. Adv.* **1991**, *9*, 217–240. [CrossRef]
4. Markets and Markets. Cell Lysis/Cell Fractionation Market-Global Forecasts to 2021. Available online: http://www.marketsandmarkets.com/Market-Reports/cell-lysis-market260138321.html (accessed on 6 March 2017).
5. Mark, D.; Haeberle, S.; Roth, G.; von Stetten, F.; Zengerle, R. Microfluidic lab-on-a-chip platforms: Requirements, characteristics and applications. *Chem. Soc. Rev.* **2010**, *39*, 1153–1182. [CrossRef] [PubMed]

6.   Andersson, H.; van den Berg, A. Microfluidic devices for cellomics: A review. *Sens. Actuators B Chem.* **2003**, *92*, 315–325. [CrossRef]

7.   Nan, L.; Jiang, Z.; Wei, X. Emerging microfluidic devices for cell lysis: A review. *Lab Chip* **2014**, *14*, 1060–1073. [CrossRef] [PubMed]

8.   Brown, R.B.; Audet, J. Current techniques for single-cell lysis. *J. R. Soc. Interface* **2008**, *5*, S131–S138. [CrossRef] [PubMed]

9.   Mahalanabis, M.; Al-Muayad, H.; Kulinski, M.D.; Altman, D.; Klapperich, C.M. Cell lysis and DNA extraction of gram-positive and gram-negative bacteria from whole blood in a disposable microfluidic chip. *Lab Chip* **2009**, *9*, 2811–2817. [CrossRef] [PubMed]

10. Engler, C.R. Disruption of microbial cells. In *Comnrehensive Biotechnoloy*; Pergamon Press: Oxford, UK, 1985; Volume 2, pp. 305–324.

11. Hammond, S.M.; Lambert, P.A.; Rycroft, A.N. *The Bacterial Cell Surface*; Croom Helm: London, UK, 1984.

12. Ghuysen, J.-M. Biosynthesis of peptidoglycan. In *The Bacterial Membranes and Walls*; Springer: Dordrecht, The Netherlands, 1973; pp. 37–130.

13. McIntosh, H.M. An Ultrastructural Study of Poly-/3-Hydroxybutyrate Separation from Alcaligenes Eutrophus by Selective Envelope Degradation. Ph.D. Thesis, University of York, York, UK, 1989.

14. Madigan, M.T.; Martinko, J.M.; Dunlap, P.V.; Clark, D.P. Brock biology of microorganisms 12th edn. *Int. Microbiol.* **2008**, *11*, 65–73.

15. Grayson, P.; Evilevitch, A.; Inamdar, M.M.; Purohit, P.K.; Gelbart, W.M.; Knobler, C.M.; Phillips, R. The effect of genome length on ejection forces in bacteriophage lambda. *Virology* **2006**, *348*, 430–436. [CrossRef] [PubMed]

16. Silhavy, T.J.; Kahne, D.; Walker, S. The bacterial cell envelope. *Cold Spring Harb. Perspect. Biol.* **2010**, *2*, a000414. [CrossRef] [PubMed]

17. Engler, C.R.; Robinson, C.W. Disruption of candida utilis cells in high pressure flow devices. *Biotechnol. Bioeng.* **1981**, *23*, 765–780. [CrossRef]

18. Middelberg, A.P.J. 2 microbial cell disruption by high-pressure homogenization. In *Downstream Processing of Proteins: Methods and Protocols*; Humana Press: New York, NY, USA, 2000; pp. 11–21.

19. Sauer, T.; Robinson, C.W.; Glick, B.R. Disruption of native and recombinant *Escherichia coli* in a high-pressure homogenizer. *Biotechnol. Bioeng.* **1989**, *33*, 1330–1342. [CrossRef] [PubMed]

20. Augenstein, D.C.; Thrasher, K.; Sinskey, A.J.; Wang, D.I.C. Optimization in the recovery of a labile intracellular enzyme. *Biotechnol. Bioeng.* **1974**, *16*, 1433–1447. [CrossRef] [PubMed]

21. Schütte, H.; Kroner, K.H.; Hustedt, H.; Kula, M.R. Experiences with a 20 litre industrial bead mill for the disruption of microorganisms. *Enzym. Microb. Technol.* **1983**, *5*, 143–148. [CrossRef]

22. Chisti, Y.; Moo-Young, M. Disruption of microbial cells for intracellular products. *Enzym. Microb. Technol.* **1986**, *8*, 194–204. [CrossRef]

23. Taskova, R.M.; Zorn, H.; Krings, U.; Bouws, H.; Berger, R.G. A comparison of cell wall disruption techniques for the isolation of intracellular metabolites from pleurotus and lepista sp. *Z. Naturforsch. C J. Biosci.* **2006**, *61*, 347–350. [CrossRef]

24. Ho, C.W.; Tan, W.S.; Yap, W.B.; Ling, T.C.; Tey, B.T. Comparative evaluation of different cell disruption methods for the release of recombinant hepatitis b core antigen from *Escherichia coli*. *Biotechnol. Bioprocess Eng.* **2008**, *13*, 577–583. [CrossRef]

25. Goldberg, S. Mechanical/physical methods of cell disruption and tissue homogenization. *Methods Mol. Biol.* **2008**, *424*, 3–22. [PubMed]

26. Johnson, B.H.; Hecht, M.H. Cells by repeated cycles of freezing and thawing. *Biotechnology* **1994**, *12*, 1357. [CrossRef] [PubMed]

27. Watson, J.S.; Cumming, R.H.; Street, G.; Tuffnell, J.M. *Release of Intracellular Protein by Thermolysis*; Ellis Horwood: London, UK, 1987; pp. 105–109.

28. Wang, B.; Merva, M.; Williams, W.V.; Weiner, D.B. Large-scale preparation of plasmid DNA by microwave lysis. *Biotechniques* **1995**, *18*, 554. [PubMed]

29. Zhu, K.; Jin, H.; He, Z.; Zhu, Q.; Wang, B. A continuous method for the large-scale extraction of plasmid DNA by modified boiling lysis. *Nat. Protoc.* **2007**, *1*, 3088–3093. [CrossRef] [PubMed]

30. Lilly, M.D.; Dunnill, P. Isolation of intracellular enzymes from micro-organisms-the development of a continuous process. In *Fermentation Advances*; Academic Press: London, UK, 1969; pp. 225–248.

31. Balasundaram, B.; Harrison, S.; Bracewell, D.G. Advances in product release strategies and impact on bioprocess design. *Trends Biotechnol.* **2009**, *27*, 477–485. [CrossRef] [PubMed]
32. Lee, A.K.; Lewis, D.M.; Ashman, P.J. Microalgal cell disruption by hydrodynamic cavitation for the production of biofuels. *J. Appl. Phycol.* **2015**, *27*, 1881–1889. [CrossRef]
33. Capocellia, M.; Prisciandarob, M.; Lanciac, A.; Musmarraa, D. Comparison between hydrodynamic and acoustic cavitation in microbial cell disruption. *Chem. Eng.* **2014**, *38*. [CrossRef]
34. Fonseca, L.P.; Cabral, J. Penicillin acylase release from *Escherichia coli* cells by mechanical cell disruption and permeabilization. *J. Chem. Technol. Biotechnol.* **2002**, *77*, 159–167. [CrossRef]
35. Chen, Y.-C.; Chen, L.-A.; Chen, S.-J.; Chang, M.-C.; Chen, T.-L. A modified osmotic shock for periplasmic release of a recombinant creatinase from *Escherichia coli*. *Biochem. Eng. J.* **2004**, *19*, 211–215. [CrossRef]
36. Byreddy, A.R.; Gupta, A.; Barrow, C.J.; Puri, M. Comparison of cell disruption methods for improving lipid extraction from thraustochytrid strains. *Mar. Drugs* **2015**, *13*, 5111–5127. [CrossRef] [PubMed]
37. Bimboim, H.C.; Doly, J. A rapid alkaline extraction procedure for screening recombinant plasmid DNA. *Nucleic Acids Res.* **1979**, *7*, 1513–1523. [CrossRef]
38. Stanbury, P.F.; Whitaker, A. *Principles of Fermentation Technology*; Pergamon Press: Oxford, UK, 1984.
39. Tamura, K.; Aotsuka, T. Rapid isolation method of animal mitochondrial DNA by the alkaline lysis procedure. *Biochem. Genet.* **1988**, *26*, 815–819. [CrossRef] [PubMed]
40. Feliciello, I.; Chinali, G. A modified alkaline lysis method for the preparation of highly purified plasmid DNA from *Escherichia coli*. *Anal. Biochem.* **1993**, *212*, 394–401. [CrossRef] [PubMed]
41. Sharma, R.; Dill, B.D.; Chourey, K.; Shah, M.; VerBerkmoes, N.C.; Hettich, R.L. Coupling a detergent lysis/cleanup methodology with intact protein fractionation for enhanced proteome characterization. *J. Proteome Res.* **2012**, *11*, 6008–6018. [CrossRef] [PubMed]
42. PanReac AppliChem. Detergents—More than Foam! Available online: https://www.applichem.com/en/literature/brochures/brochures-biochemical-support/detergents/ (accessed on 6 March 2017).
43. Andrews, B.A.; Asenjo, J.A. Enzymatic lysis and disruption of microbial cells. *Trends Biotechnol.* **1987**, *5*, 273–277. [CrossRef]
44. Salazar, O.; Asenjo, J.A. Enzymatic lysis of microbial cells. *Biotechnol. Lett.* **2007**, *29*, 985–994. [CrossRef] [PubMed]
45. Anand, H.; Balasundaram, B.; Pandit, A.B.; Harrison, S.T.L. The effect of chemical pretreatment combined with mechanical disruption on the extent of disruption and release of intracellular protein from *E. coli*. *Biochem. Eng. J.* **2007**, *35*, 166–173. [CrossRef]
46. Beebe, D.J.; Mensing, G.A.; Walker, G.M. Physics and applications of microfluidics in biology. *Annu. Rev. Biomed. Eng.* **2002**, *4*, 261–286. [CrossRef] [PubMed]
47. Khandurina, J.; McKnight, T.E.; Jacobson, S.C.; Waters, L.C.; Foote, R.S.; Ramsey, J.M. Integrated system for rapid PCR-based DNA analysis in microfluidic devices. *Anal. Chem.* **2000**, *72*, 2995–3000. [CrossRef] [PubMed]
48. Burns, M.A.; Johnson, B.N.; Brahmasandra, S.N.; Handique, K.; Webster, J.R.; Krishnan, M.; Sammarco, T.S.; Man, P.M.; Jones, D.; Heldsinger, D. An integrated nanoliter DNA analysis device. *Science* **1998**, *282*, 484–487. [CrossRef] [PubMed]
49. Lin, Z.; Cai, Z. Cell lysis methods for high-throughput screening or miniaturized assays. *Biotechnol. J.* **2009**, *4*, 210–215. [CrossRef] [PubMed]
50. Berasaluce, A.; Matthys, L.; Mujika, J.; Antoñana-Díez, M.; Valero, A.; Agirregabiria, M. Bead beating-based continuous flow cell lysis in a microfluidic device. *RSC Adv.* **2015**, *5*, 22350–22355. [CrossRef]
51. Pham, V.T.H.; Truong, V.K.; Mainwaring, D.E.; Guo, Y.; Baulin, V.A.; Al Kobaisi, M.; Gervinskas, G.; Juodkazis, S.; Zeng, W.R.; Doran, P.P. Nanotopography as a trigger for the microscale, autogenous and passive lysis of erythrocytes. *J. Mater. Chem. B* **2014**, *2*, 2819–2826. [CrossRef]
52. Di Carlo, D.; Jeong, K.H.; Lee, L.P. Reagentless mechanical cell lysis by nanoscale barbs in microchannels for sample preparation. *Lab Chip* **2003**, *3*, 287–291. [CrossRef] [PubMed]
53. Kido, H.; Micic, M.; Smith, D.; Zoval, J.; Norton, J.; Madou, M. A novel, compact disk-like centrifugal microfluidics system for cell lysis and sample homogenization. *Colloids Surf. B Biointerfaces* **2007**, *58*, 44–51. [CrossRef] [PubMed]
54. Kim, J.; Hee Jang, S.; Jia, G.; Zoval, J.V.; Da Silva, N.A.; Madou, M.J. Cell lysis on a microfluidic CD (compact disc). *Lab Chip* **2004**, *4*, 516–522. [CrossRef] [PubMed]

55. Madou, M.; Zoval, J.; Jia, G.; Kido, H.; Kim, J.; Kim, N. Lab on a CD. *Annu. Rev. Biomed. Eng.* **2006**, *8*, 601–628. [CrossRef] [PubMed]
56. Tsougeni, K.; Papadakis, G.; Gianneli, M.; Grammoustianou, A.; Constantoudis, V.; Dupuy, B.; Petrou, P.S.; Kakabakos, S.E.; Tserepi, A.; Gizeli, E. Plasma nanotextured polymeric lab-on-a-chip for highly efficient bacteria capture and lysis. *Lab Chip* **2016**, *16*, 120–131. [CrossRef] [PubMed]
57. Kim, J.; Johnson, M.; Hill, P.; Gale, B.K. Microfluidic sample preparation: Cell lysis and nucleic acid purification. *Integr. Biol. Quant. Biosci. Nano Macro* **2009**, *1*, 574–586. [CrossRef] [PubMed]
58. Yeung, S.-W.; Lee, T.M.-H.; Cai, H.; Hsing, I.M. A DNA biochip for on-the-spot multiplexed pathogen identification. *Nucleic Acids Res.* **2006**, *34*, e118. [CrossRef] [PubMed]
59. Liu, R.H.; Yang, J.; Lenigk, R.; Bonanno, J.; Grodzinski, P. Self-contained, fully integrated biochip for sample preparation, polymerase chain reaction amplification, and DNA microarray detection. *Anal. Chem.* **2004**, *76*, 1824–1831. [CrossRef] [PubMed]
60. Buser, J.R.; Zhang, X.; Byrnes, S.A.; Ladd, P.D.; Heiniger, E.K.; Wheeler, M.D.; Bishop, J.D.; Englund, J.A.; Lutz, B.; Weigl, B.H. A disposable chemical heater and dry enzyme preparation for lysis and extraction of DNA and RNA from microorganisms. *Anal. Methods* **2016**, *8*, 2880–2886. [CrossRef]
61. Kashyap, A.; Autebert, J.; Delamarche, E.; Kaigala, G.V. Selective local lysis and sampling of live cells for nucleic acid analysis using a microfluidic probe. *Sci. Rep.* **2016**, *6*, 29579. [CrossRef] [PubMed]
62. Hall, J.A.; Felnagle, E.; Fries, M.; Spearing, S.; Monaco, L.; Steele, A. Evaluation of cell lysis procedures and use of a micro fluidic system for an automated DNA-based cell identification in interplanetary missions. *Planet. Space Sci.* **2006**, *54*, 1600–1611. [CrossRef]
63. Kim, Y.-B.; Park, J.-H.; Chang, W.-J.; Koo, Y.-M.; Kim, E.-K.; Kim, J.-H. Statistical optimization of the lysis agents for gram-negative bacterial cells in a microfluidic device. *Biotechnol. Bioprocess Eng.* **2006**, *11*, 288–292. [CrossRef]
64. Heo, J.; Thomas, K.J.; Seong, G.H.; Crooks, R.M. A microfluidic bioreactor based on hydrogel-entrapped *E. coli*: Cell viability, lysis, and intracellular enzyme reactions. *Anal. Chem.* **2003**, *75*, 22–26. [CrossRef] [PubMed]
65. Sethu, P.; Anahtar, M.; Moldawer, L.L.; Tompkins, R.G.; Toner, M. Continuous flow microfluidic device for rapid erythrocyte lysis. *Anal. Chem.* **2004**, *76*, 6247–6253. [CrossRef] [PubMed]
66. Schilling, E.A.; Kamholz, A.E.; Yager, P. Cell lysis and protein extraction in a microfluidic device with detection by a fluorogenic enzyme assay. *Anal. Chem.* **2002**, *74*, 1798–1804. [CrossRef] [PubMed]
67. Marc, P.J.; Sims, C.E.; Allbritton, N.L. Coaxial-flow system for chemical cytometry. *Anal. Chem.* **2007**, *79*, 9054–9059. [CrossRef] [PubMed]
68. Ocvirk, G.; Salimi-Moosavi, H.; Szarka, R.J.; Arriaga, E.A.; Andersson, P.E.; Smith, R.; Dovichi, N.J.; Harrison, D.J. B-galactosidase assays of single-cell lysates on a microchip: A complementary method for enzymatic analysis of single cells. *Proc. IEEE* **2004**, *92*, 115–125. [CrossRef]
69. Abolmaaty, A.; El-Shemy, M.G.; Khallaf, M.F.; Levin, R.E. Effect of lysing methods and their variables on the yield of *Escherichia coli* O157: H7 DNA and its PCR amplification. *J. Microbiol. Methods* **1998**, *34*, 133–141. [CrossRef]
70. Huang, S.-H.; Hung, L.-Y.; Lee, G.-B. Continuous nucleus extraction by optically-induced cell lysis on a batch-type microfluidic platform. *Lab Chip* **2016**, *16*, 1447–1456. [CrossRef] [PubMed]
71. Kremer, C.; Witte, C.; Neale, S.L.; Reboud, J.; Barrett, M.P.; Cooper, J.M. Shape-dependent optoelectronic cell lysis. *Angew. Chem.* **2014**, *126*, 861–865. [CrossRef]
72. Rau, K.R.; Quinto-Su, P.A.; Hellman, A.N.; Venugopalan, V. Pulsed laser microbeam-induced cell lysis: Time-resolved imaging and analysis of hydrodynamic effects. *Biophys. J.* **2006**, *91*, 317–329. [CrossRef] [PubMed]
73. Li, H.; Sims, C.E.; Wu, H.Y.; Allbritton, N.L. Spatial control of cellular measurements with the laser micropipet. *Anal. Chem.* **2001**, *73*, 4625–4631. [CrossRef] [PubMed]
74. Hellman, A.N.; Rau, K.R.; Yoon, H.H.; Venugopalan, V. Biophysical response to pulsed laser microbeam-lnduced cell lysis and molecular delivery. *J. Biophotonics* **2008**, *1*, 24–35. [CrossRef] [PubMed]
75. Quinto-Su, P.A.; Lai, H.-H.; Yoon, H.H.; Sims, C.E.; Allbritton, N.L.; Venugopalan, V. Examination of laser microbeam cell lysis in a PDMS microfluidic channel using time-resolved imaging. *Lab Chip* **2008**, *8*, 408–414. [CrossRef] [PubMed]

76. Wan, W.; Yeow, J.T. Study of a novel cell lysis method with titanium dioxide for lab-on-a-chip devices. *Biomed. Microdevices* **2011**, *13*, 527–532. [CrossRef] [PubMed]

77. Taller, D.; Richards, K.; Slouka, Z.; Senapati, S.; Hill, R.; Go, D.B.; Chang, H.-C. On-chip surface acoustic wave lysis and ion-exchange nanomembrane detection of exosomal RNA for pancreatic cancer study and diagnosis. *Lab Chip* **2015**, *15*, 1656–1666. [CrossRef] [PubMed]

78. Chen, Y.; Ding, X.; Steven Lin, S.-C.; Yang, S.; Huang, P.-H.; Nama, N.; Zhao, Y.; Nawaz, A.A.; Guo, F.; Wang, W. Tunable nanowire patterning using standing surface acoustic waves. *ACS Nano* **2013**, *7*, 3306–3314. [CrossRef] [PubMed]

79. Guo, F.; Li, P.; French, J.B.; Mao, Z.; Zhao, H.; Li, S.; Nama, N.; Fick, J.R.; Benkovic, S.J.; Huang, T.J. Controlling cell–cell interactions using surface acoustic waves. *Proc. Natl. Acad. Sci. USA* **2015**, *112*, 43–48. [CrossRef] [PubMed]

80. Marentis, T.C.; Kusler, B.; Yaralioglu, G.G.; Liu, S.; Haeggstrom, E.O.; Khuri-Yakub, B.T. Microfluidic sonicator for real-time disruption of eukaryotic cells and bacterial spores for DNA analysis. *Ultrasound Med. Biol.* **2005**, *31*, 1265–1277. [CrossRef] [PubMed]

81. Taylor, M.T.; Belgrader, P.; Furman, B.J.; Pourahmadi, F.; Kovacs, G.T.; Northrup, M.A. Lysing bacterial spores by sonication through a flexible interface in a microfluidic system. *Anal. Chem.* **2001**, *73*, 492–496. [CrossRef] [PubMed]

82. Reboud, J.; Bourquin, Y.; Wilson, R.; Pall, G.S.; Jiwaji, M.; Pitt, A.R.; Graham, A.; Waters, A.P.; Cooper, J.M. Shaping acoustic fields as a toolset for microfluidic manipulations in diagnostic technologies. *Proc. Natl. Acad. Sci. USA* **2012**, *109*, 15162–15167. [CrossRef] [PubMed]

83. Wei, X.; Nan, L.; Ren, J.; Liu, X.; Jiang, Z. Surface acoustic wave induced thermal lysis of red blood cells in microfluidic channel. In Proceedings of the 19th International Conference on Miniaturized Systems for Chemistry and Life Sciences, Gyeongju, Korea, 25–29 October 2015.

84. Tandiono, T.; Ow, D.S.; Driessen, L.; Chin, C.S.; Klaseboer, E.; Choo, A.B.; Ohl, S.W.; Ohl, C.D. Sonolysis of *Escherichia coli* and pichia pastoris in microfluidics. *Lab Chip* **2012**, *12*, 780–786. [CrossRef] [PubMed]

85. Zhang, H.; Jin, W. Determination of different forms of human interferon-gamma in single natural killer cells by capillary electrophoresis with on-capillary immunoreaction and laser-induced fluorescence detection. *Electrophoresis* **2004**, *25*, 1090–1095. [CrossRef] [PubMed]

86. Ohshima, T.; Sato, M.; Saito, M. Selective release of intracellular protein using pulsed electric field. *J. Electrost.* **1995**, *35*, 103–112. [CrossRef]

87. Ameri, S.K.; Singh, P.K.; Dokmeci, M.R.; Khademhosseini, A.; Xu, Q.; Sonkusale, S.R. All electronic approach for high-throughput cell trapping and lysis with electrical impedance monitoring. *Biosens. Bioelectron.* **2014**, *54*, 462–467. [CrossRef] [PubMed]

88. Jiang, F.; Chen, J.; Yu, J. Design and application of a microfluidic cell lysis microelectrode chip. *Instrum. Sci. Technol.* **2016**, *44*, 223–232. [CrossRef]

89. De Lange, N.; Tran, T.M.; Abate, A.R. Electrical lysis of cells for detergent-free droplet assays. *Biomicrofluidics* **2016**, *10*, 024114. [CrossRef] [PubMed]

90. Escobedo, C.; Bürgel, S.C.; Kemmerling, S.; Sauter, N.; Braun, T.; Hierlemann, A. On-chip lysis of mammalian cells through a handheld corona device. *Lab Chip* **2015**, *15*, 2990–2997. [CrossRef] [PubMed]

91. Besant, J.D.; Das, J.; Sargent, E.H.; Kelley, S.O. Proximal bacterial lysis and detection in nanoliter wells using electrochemistry. *ACS Nano* **2013**, *7*, 8183–8189. [CrossRef] [PubMed]

92. Gabardo, C.M.; Kwong, A.M.; Soleymani, L. Rapidly prototyped multi-scale electrodes to minimize the voltage requirements for bacterial cell lysis. *Analyst* **2015**, *140*, 1599–1608. [CrossRef] [PubMed]

93. Li, X.; Zhao, S.; Hu, H.; Liu, Y.-M. A microchip electrophoresis-mass spectrometric platform with double cell lysis nano-electrodes for automated single cell analysis. *J. Chromatogr. A* **2016**, *1451*, 156–163. [CrossRef] [PubMed]

94. Wassermann, K.J.; Maier, T.; Keplinger, F.; Peham, J.R. A novel sample preparation concept for sepsis diagnostics using high frequency electric fields. In Proceedings of the 1st World Congress on Electroporation and Pulsed Electric Fields in Biology, Medicine and Food & Environmental Technologies, Portorož, Slovenia, 6–10 September 2015; Jarm, T., Kramar, P., Eds.; Springer: Singapore, 2016; pp. 302–306.

95. Ma, S.; Bryson, B.D.; Sun, C.; Fortune, S.M.; Lu, C. RNA extraction from a mycobacterium under ultrahigh electric field intensity in a microfluidic device. *Anal. Chem.* **2016**, *88*, 5053–5057. [CrossRef] [PubMed]

96. Islam, M.I.; Kuryllo, K.; Selvaganapathy, P.R.; Li, Y.; Deen, M.J. A microfluidic sample preparation device for pre-concentration and cell lysis using a nanoporous membrane. In Proceedings of the 17th International Conference on Miniaturized Systems for Chemistry and Life Sciences, Freiburg, Germany, 27–31 October 2013.

97. Sedgwick, H.; Caron, F.; Monaghan, P.B.; Kolch, W.; Cooper, J.M. Lab-on-a-chip technologies for proteomic analysis from isolated cells. *J. R. Soc. Interface* **2008**, *5*, S123. [CrossRef] [PubMed]

98. Lim, J.K.; Zhou, H.; Tilton, R.D. Liposome rupture and contents release over coplanar microelectrode arrays. *J. Colloid Interface Sci.* **2009**, *332*, 113–121. [CrossRef] [PubMed]

99. Lu, K.Y.; Wo, A.M.; Lo, Y.J.; Chen, K.C.; Lin, C.M.; Yang, C.R. Three dimensional electrode array for cell lysis via electroporation. *Biosens. Bioelectron.* **2006**, *22*, 568–574. [CrossRef] [PubMed]

100. Church, C.; Zhu, J.; Huang, G.; Tzeng, T.-R.; Xuan, X. Integrated electrical concentration and lysis of cells in a microfluidic chip. *Biomicrofluidics* **2010**, *4*, 044101. [CrossRef] [PubMed]

101. Lee, S.W.; Yowanto, H.; Tai, Y.C. A micro cell lysis device. In Proceedings of the IEEE Eleventh Annual International Workshop on Micro Electro Mechanical Systems (MEMS 98). An Investigation of Micro Structures, Sensors, Actuators, Machines and Systems (Cat. No. 98CH36176), Heidelberg, Germany, 25–29 January 1998; pp. 443–447.

102. Lee, D.W.; Cho, Y.-H. A continuous electrical cell lysis device using a low dc voltage for a cell transport and rupture. *Sens. Actuators B Chem.* **2007**, *124*, 84–89. [CrossRef]

103. Wang, H.Y.; Lu, C. Electroporation of mammalian cells in a microfluidic channel with geometric variation. *Anal. Chem.* **2006**, *78*, 5158–5164. [CrossRef] [PubMed]

104. Lu, H.; Schmidt, M.A.; Jensen, K.F. A microfluidic electroporation device for cell lysis. *Lab Chip* **2005**, *5*, 23–29. [CrossRef] [PubMed]

105. De la Rosa, C.; Kaler, K.V. Electro-disruption of *Escherichia coli* bacterial cells on a microfabricated chip. *Conf. Proc. IEEE Eng. Med. Biol. Soc.* **2006**, *1*, 4096–4099. [PubMed]

106. Wang, H.Y.; Lu, C. Microfluidic chemical cytometry based on modulation of local field strength. *Chem. Commun.* **2006**, 3528–3530. [CrossRef] [PubMed]

107. Wang, H.Y.; Bhunia, A.K.; Lu, C. A microfluidic flow-through device for high throughput electrical lysis of bacterial cells based on continuous dc voltage. *Biosens. Bioelectron.* **2006**, *22*, 582–588. [CrossRef] [PubMed]

108. Bao, N.; Lu, C. A microfluidic device for physical trapping and electrical lysis of bacterial cells. *Appl. Phys. Lett.* **2008**, *92*, 214103. [CrossRef]

109. Svec, D.; Andersson, D.; Pekny, M.; Sjöback, R.; Kubista, M.; Ståhlberg, A. Direct cell lysis for single-cell gene expression profiling. *Front. Oncol.* **2013**, *3*, 274. [CrossRef] [PubMed]

110. Kemmerling, S.; Arnold, S.A.; Bircher, B.A.; Sauter, N.; Escobedo, C.; Dernick, G.; Hierlemann, A.; Stahlberg, H.; Braun, T. Single-cell lysis for visual analysis by electron microscopy. *J. Struct. Biol.* **2013**, *183*, 467–473. [CrossRef] [PubMed]

micromachines

MDPI

Article

Cell Migration According to Shape of Graphene Oxide Micropatterns

Sung Eun Kim [1,†], Min Sung Kim [2,†], Yong Cheol Shin [1], Seong Un Eom [1], Jong Ho Lee [1], Dong-Myeong Shin [3], Suck Won Hong [1], Bongju Kim [4], Jong-Chul Park [2], Bo Sung Shin [1], Dohyung Lim [5,*] and Dong-Wook Han [1,*]

[1]   Department of Cogno-Mechatronics Engineering, College of Nanoscience and Nanotechnology, Pusan National University, Busan 46241, Korea; 01048470363@naver.com (S.E.K.); choel15@naver.com (Y.C.S.); sueom89@gmail.com (S.U.E.); pignunssob@naver.com (J.H.L.); swhong@pusan.ac.kr (S.W.H.); bosung@pusan.ac.kr (B.S.S.)
[2]   Cellbiocontrol Laboratory, Department of Medical Engineering, Yonsei University College of Medicine, Seoul 03722, Korea; kimminsec@nate.com (M.S.K.); parkjc@yuhs.ac (J.-C.P.)
[3]   Research Center for Energy Convergence Technology, Pusan National University, Busan 46241, Korea; dmshin@pusan.ac.kr
[4]   Dental Life Science Research Institute, Seoul National University Dental Hospital, Seoul 03080, Korea; bjkim016@gmail.com
[5]   Department of Mechanical Engineering, Sejong University, Seoul 05006, Korea
*   Correspondence: dli349@sejong.ac.kr (D.L.); nanohan@pusan.ac.kr (D.-W.H.); Tel.: +82-2-3408-4333 (D.L.); +82-51-510-7725 (D.-W.H.)
†   These authors contributed equally to this work.

Academic Editors: Chang-Hwan Choi, Aaron T. Ohta and Wenqi Hu
Received: 26 August 2016; Accepted: 7 October 2016; Published: 14 October 2016

Abstract: Photolithography is a unique process that can effectively manufacture micro/nano-sized patterns on various substrates. On the other hand, the meniscus-dragging deposition (MDD) process can produce a uniform surface of the substrate. Graphene oxide (GO) is the oxidized form of graphene that has high hydrophilicity and protein absorption. It is widely used in biomedical fields such as drug delivery, regenerative medicine, and tissue engineering. Herein, we fabricated uniform GO micropatterns via MDD and photolithography. The physicochemical properties of the GO micropatterns were characterized by atomic force microscopy (AFM), scanning electron microscopy (SEM), and Raman spectroscopy. Furthermore, cell migration on the GO micropatterns was investigated, and the difference in cell migration on triangle and square GO micropatterns was examined for their effects on cell migration. Our results demonstrated that the GO micropatterns with a desired shape can be finely fabricated via MDD and photolithography. Moreover, it was revealed that the shape of GO micropatterns plays a crucial role in cell migration distance, speed, and directionality. Therefore, our findings suggest that the GO micropatterns can serve as a promising biofunctional platform and cell-guiding substrate for applications to bioelectric devices, cell-on-a-chip, and tissue engineering scaffolds.

Keywords: photolithography; meniscus-dragging deposition; graphene oxide; micropatterns; cell migration

---

1. Introduction

Photolithography is a unique process that can facilitate the manufacturing of micro-sized patterns on various substrates. This process has significantly higher resolution than other patterning methods as well as good reproducibility and efficiency from an economical and temporal perspective.

Therefore, photolithography has been widely used as a patterning process to fabricate semi-conductors, stretchable devices, and medical devices [1,2].

Graphene oxide (GO), the oxidized form of graphene, is a carbon-based hexagonal structure with oxygen containing groups such as carboxyl, hydroxyl, and epoxy groups [3–5]. GO has good dispersion in aqueous solutions, which is useful for uniformly coating GO [6]. In addition, GO presents an open surface for noncovalent interactions with biomolecules. Recent research has shown that GO can enhance cellular behaviors including attachment, proliferation, and differentiation due to various functional groups on its surface that can promote cellular behaviors through interactions with cells [7–10]. Based on this reason, it is inferred that GO-coated substrates can induce aligned array of cells. GO-coated substrates can be fabricated by various coating techniques, including filtration/transfer-based film formation, spin coating, air-spraying, dip coating, Langmuir–Blodgett deposition, and wire-wound rod coating, for electrical devices and medical applications. Some of these methods produce relatively non-uniform thin films because of the aggregation of GO particles. In addition, the majority of these techniques for the production of GO is not easy to scale over a large area. Recently, the meniscus-dragging deposition (MDD) technique, which is a microliter-scale solution process for fabricating thin film-coated substrates with a significant decrease of the solution consumption, has been spotlighted because the process can easily and uniformly fabricate GO-coated substrates. The MDD technique can develop highly uniform GO films on substrates by dragging the meniscus of a GO suspension trapped between a deposition plate and a coating substrate in an alternating back-and-forth motion [11–14]. Therefore, in this study, we fabricated GO micropatterns on a glass substrate via MDD and photolithography techniques and investigated their effects on cell migration.

Directional cell migration is critical for many important biological processes, including angiogenesis, tumor metastasis, wound healing, and nerve regeneration. Most work on the directional control of cell motility has focused on the role of gradients of motility factors such as platelet-derived growth factor, fibroblast growth factor, and epidermal growth factor, with the general concept that cells physically move up the gradient of a soluble attractant. These factors promote cell migration by activating members of the Rho family of GTPases-Rac and CDC43, which induce the formation of actin-based lamellipodia, filopodia, and fascin-containing microspikes that drive cell extension. On the other hand, recently, many studies have been concerned on the development of micropatterns that can induce cell migration to a desired direction. The specific micropatterns can induce and guide the cell migration by providing physical and topographical cues. In addition, the micropatterned substrates are effectively and consistently able to provide guidance cues [15–17]. Therefore, we speculated that the GO micropatterns might guide the cell migration [18–21].

Herein, we fabricated the GO micropatterns on a slide glass evenly using MDD and photolithography techniques. In addition, L-929 fibroblasts were cultured on the GO micropatterns to explore the effects of the GO micropatterns on cell migration. Furthermore, the difference in cell migration according to the pattern shape was investigated to explore the potential of GO micropatterns as a biofunctional platform for bioelectric devices and tissue engineering applications.

2. Materials and Methods

2.1. Preparation of GO-Coated Substrates and GO Micropatterns

A GO solution was purchased from Sigma-Aldrich Co. (St. Louis, MO, USA). To prevent the defects of GO during photolithography, 25 mm × 75 mm slide glass was pre-treated by placing it into a piranha solution (H_2SO_4:H_2O_2 = 3:1) for 30 min. The deposition plate was placed on the coating substrate at an angle of 30°. The 120-μL GO solution (4 mg/mL in distilled water) was injected into the wedge between the plate and slide glass. The deposition plate was moved linearly in a back-and-forth motion at a constant speed of 15 mm/s in a 35% humidified atmosphere to deposit the GO on the substrate. After the coating process, GO-coated slide glass was dried in a vacuum oven at 80 °C for 30 min.

Positive photoresists (PRs, az5214e) were spin-coated on the GO-coated slide glass and soft baked at 95 °C for 5 min. Next, the substrates were exposed to 20 mW of ultra-violet (UV) lights for 6 s through a micropatterned chrome mask. During the developing step, the exposed PRs were dissolved by a developer (AZ 300 MIF, AZ Electronic Materials, Branchburg, NJ, USA). Then, 100 sccm of O_2 plasma was applied to the remaining GO between the PRs and the slide glass for 6 min. After all of these steps, the substrates were washed with acetone and dried under N_2 gas to remove the PRs.

2.2. Physicochemical Characterization of GO Micropatterns

The topography of the GO-coated slide glass using the MDD method was characterized by atomic force microscopy (AFM, NX10, Park Systems Co., Suwon, Korea) in air at room temperature (RT). Imaging was performed in non-contact mode with a Multi 75 silicon scanning probe at a resonant frequency of ~300 kHz. Image analysis was performed using XEI Software (version 1.7.1, Park Systems Co.). To examine the morphology of GO micropatterns, the GO micropatterns on substrates were observed with a field emission scanning electron microscope (FESEM, Hitachi S-4700, Tokyo, Japan) at an accelerating voltage of 5 kV. Compositional analysis of the GO micropatterns was performed via Raman spectroscopy (Micro Raman PL Mapping System, Dongwoo Optron Co., Ltd., Kwangju-si, Korea) with excitation at 532 nm using an Ar-ion laser with a radiant power of 5 mW at RT.

2.3. Time-Lapse Imaging and Analysis of Cell Migration on GO Micropatterns

L-929 fibroblasts were purchased from the American Type Culture Collection (ATCC, Rockville, MD, USA) and routinely maintained in Dulbecco's modified Eagle's medium (Welgene, Daegu, Korea) supplemented with 10% fetal bovine serum (Welgene) and 1% antibiotic–antimycotic solution (including 10,000 units of penicillin, 10 mg of streptomycin, and 25 µg of amphotericin B per mL, Sigma-Aldrich Co.) at 37 °C in a humidified atmosphere containing 5% CO_2.

Cell migration images were captured with an Olympus IX81 inverted fluorescence microscope (Olympus Optical Co., Osaka, Japan). Captured images were imported into ImageJ (ImageJ, version 1.37 by Wayne Rasband, National Institutes of Health, Bethesda, MD, USA), and image analysis was carried out with the manual tracking and chemotaxis tool plug-in (version 1.01, distributed by ibidi GmbH, Munchen, Germany) [22–25]. The *XY* coordinates of each cell were obtained using the manual tracking plug-in in the ImageJ program. Only one cell of each group was tracked and the center of each cell was tracked 3 times to obtain accuracy. The tracked data were imported into the chemotaxis plug-in. The cell migration speed was computed automatically, and the cell migration pathway was plotted with the chemotaxis tool. The migration speed indicates how fast cells move in response to the stimulation, calculated using the total length of the migration path divided by the total observation time. The distance is the total length of the cell migration path during the observation time. Cells undergoing division, death, or migration outside the field of view were excluded from the analysis.

2.4. Statistical Analysis

All variables were tested in three independent cultures for each in vitro experiment, which was repeated twice ($n = 6$). The quantitative data are given as the mean ± standard deviation (SD). A one-way analysis of variance (ANOVA, SAS Institute Inc., Cary, NC, USA) was performed to analyze the difference in cell migration according to the pattern shape by a Tukey's honestly significant difference (HSD) test. A value of $p < 0.05$ was considered statistically significant.

3. Results and Discussion

3.1. Preparation of GO-Coated Substrates and GO Micropatterns

The procedures of fabrication of GO micropatterns on a slide glass are divided into two main steps (Figure 1). First, to produce uniform GO layers on the substrate, the MDD method was used. Figure 2a displays a schematic illustration of the MDD process, which can allow GO particles to be uniformly coated on the substrate. In brief, the deposition plate was moved at a constant angle and speed to deposit the GO solution on the substrate. Then, there were differences in evaporation ratio because of the meniscus phenomena, and the GO particles were evenly coated on the surface of substrate.

Figure 2b displays digital photographs of GO-coated slide glass with different pre-treatments. The piranha-treated slide glass was coated uniformly. However, when the slide glass was pre-treated with octadecyltrichlorosilane (OTS), GO was coated non-uniformly. This could be in part due to the hydrophobicity of OTS. This result confirmed that the hydrophilic surface of slide glass is more suitable for GO coating than a hydrophobic surface [26,27]. We chose the piranha-treated slide glass for the manufacturing of uniform GO micropatterns. Figure 2c displays the AFM images of GO-coated slide glass surface with different methods of GO coating. The surface of the GO-coated slide glass using the MDD method was formed uniformly on slide glass and had a lower average of surface roughness than the GO-coated slide glass using drop-casting. Lower roughness means that the GO was highly uniformly coated on the slide glass without an aggregation of GO particles. Therefore, it is indicated that the MDD method with proper conditions is a suitable procedure for manufacturing uniformly GO-coated slide glass.

The second step of photolithography is PR coating and the developing process (Figure 1). A PR is a photopolymer resin that can regulate the cross-linking between the molecules via light energy. After PR coating, the light source was passed through a chrome mask that carved specific micropatterns. Then, the cross-linked bonds between PRs became weak, and the PRs could be dissolved in the developer solution.

Finally, the substrate containing the GO layers and the micropatterned PR was exposed to O_2 plasma to form GO micropatterns because the GO layers under the micropatterned PR did not react with the plasma due to the protection of the PR. After washing with acetone and drying under N_2 gas, the uniform GO micropatterns were obtained.

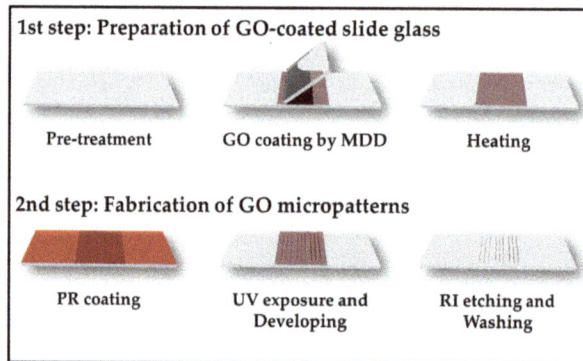

Figure 1. Schematic illustration for the preparation of graphene oxide (GO)-coated slide glass and GO micropatterns. PR: positive photoresists; MDD: meniscus-dragging deposition; UV: ultra-violet.

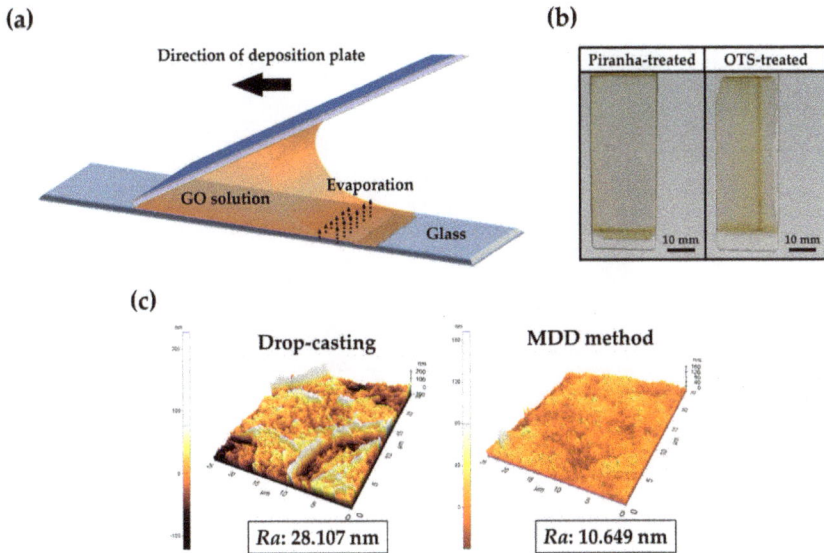

Figure 2. Preparation of GO-coated slide glass using the MDD method. (**a**) Schematic illustration of MDD technique; (**b**) Digital photographs of GO-coated slide glass in different pre-treatments; (**c**) Atomic force microscope (AFM) images and surface roughness (*Ra*) of GO-coated slide glass according to GO coating methods. OTS: octadecyltrichlorosilane.

3.2. Physicochemical Characeristics of GO Micropattenrs

Figure 3a showed the surface morphologies from FESEM. In this study, two types of GO micropatterns—square and triangle shapes—were designed on slide glass. The gaps between the square and triangle patterns were 25.7 ± 1.4 μm and 16.7 ± 0.3 μm, respectively. They were close enough for extended lamellipodia to reach into adjacent micropatterns, but far enough to momentarily confine individual cells. The side length of the square pattern was shorter than that of the triangle pattern. Although side lengths of two type patterns are different, the area of GO micropatterns is highly similar because it is important that the same quantity of GO is coated on each pattern to investigate the effects of GO.

Figure 3b displays the Raman spectra of the GO micropatterns. It was demonstrated that the spectrum of the GO micropatterns included characteristic bands of GO: the D and G bands. These characteristic peaks of GO were not shifted in the GO micropatterns. The D and G bands were observed at approximately 1390 and 1600 cm^{-1}, which were assigned to the vibration of sp^3 carbon atoms and the structural defects of the sp^2 carbon domains, respectively. In addition, in general, the intensity ratio of the D and G bands (I_D/I_G value) of GO is less than 1 because the GO has many defects on its surface [28–30]. As shown in Figure 3b, the I_D/I_G value of GO was less than 1, which is in accordance with previous studies. Therefore, it was demonstrated that the GO micropatterns were successfully formed on the slide glass.

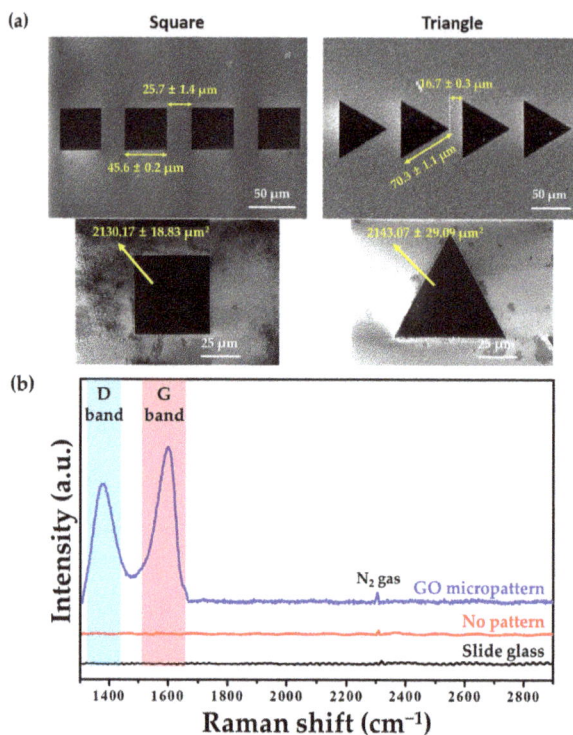

Figure 3. Physicochemical characteristics of GO micropatterns. (**a**) Field emission scanning electron microscopy (FESEM) images and (**b**) Raman spectra of GO micropatterns. Characteristic bands of GO including D and G bands were observed in GO micropatterns.

3.3. Effects of GO Micropatterns on Cell Migration

The migration of L-929 fibroblasts on GO micropatterns was investigated via optical microscopy (Figure 4). We found that, firstly, L-929 fibroblasts moved on micropatterns rather than the unpatterned slide glass region due to the GO. This can be attributed the fact that the functional groups of the GO surface can provide a favorable environment for cell attachment and growth. In a previous study, it was found that GO can regulate cellular responses and attract the cells because GO has many hydrophilic functional groups including hydroxyl, carboxyl, and epoxy groups [3,4,7,9]. As a result, the interactions between GO and cells can be promoted via the hydrophilic functional groups of the GO surface, which results in successful cell adhesion on the GO micropatterns.

The cells on GO micropatterns gradually migrated following the GO micropatterns. As shown in Figure 4a, the L-929 fibroblasts on the triangle micropatterns migrated from left to right initially, and then moved backwards to the left hand side. In addition, interestingly, the cells moved along an oblique side of the triangle micropatterns, toward the vertex of the triangle patterns continuously. It has been revealed that the asymmetric relative positioning of the micropatterns can provide both path and directionality. A previous study has demonstrated that lamellipodia attachment has an influence on the shape of substrate [31,32]. It was found that the migration distance of the cells on the symmetric square micropatterns was significantly ($p < 0.05$) shorter than that on the triangle micropatterns (Figure 4b), although the cells on the square micropatterns also migrated to the next micropattern (Videos S1 and S2).

GO micropattern

(a) triangle   (b) square

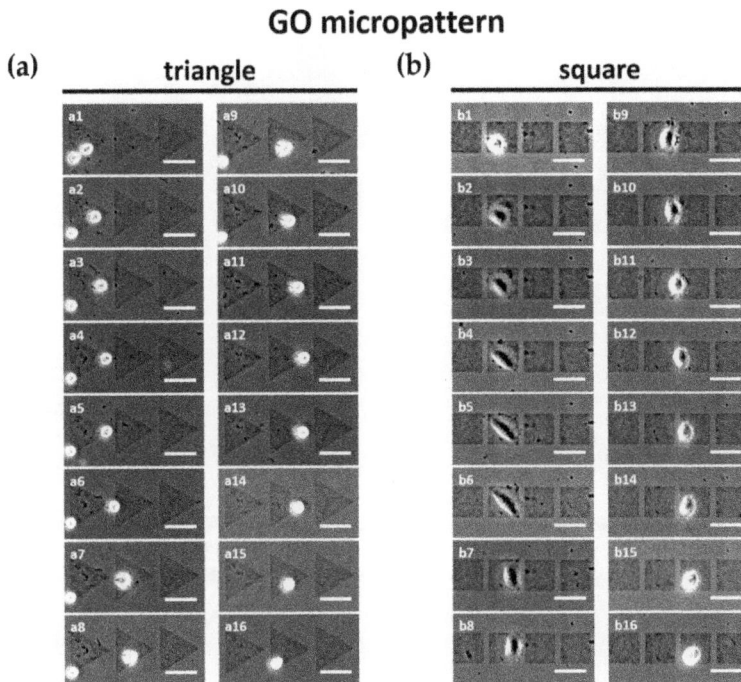

Figure 4. Time-lapse images of L-929 fibroblasts on (a) triangle and (b) square GO micropatterns for 12 h. Scale bars are 50 μm. (a) L-929 fibroblasts on triangle GO micropatterns moved from left to right initially (a1–a8), and then moved backwards to the left hand side (a9–a16); (b) The migration distance of L-929 fibroblasts on square GO micropatterns was significantly shorter than that on the triangle GO micropatterns (b1–b16).

To quantitatively analyze the migration of the L-929 fibroblasts on the GO micropatterns, trajectories, migration distance, and average migration speed were calculated. Figure 5a presents the trajectories of the cells on each micropattern for 12 h. It is demonstrated that cells on the triangle micropatterns tended to move along an oblique side of the triangle GO micropatterns. This can be partly explained by the fact that the lamellipodia of cells tends to reach the edge of the shape or the sharp part [33,34]. Furthermore, as shown in Figure 5b,c, both migration distance and speed of the cells on the triangle micropatterns were significantly ($p < 0.05$) higher than those on the square micropatterns. In addition, it is related to a change of cell morphology depending on the topography of the patterns [31,32]. In Figure 4, the cell morphologies on the triangle and square patterns are apparently different. The cell morphology on the square patterns was more spread on the GO micropatterns than that on the triangle patterns. This could also affect the slower speed and shorter distance of the cell migration on the square patterns.

Taken together, our results demonstrated that the cell migration was strongly dependent on the shape of the GO micropatterns, and the triangle GO micropatterns were more suitable for enhancing cell migration in terms of migration distance, speed, and directionality. Consequently, it is suggested that the GO micropatterns with specific geometrical cues can consistently regulate and guide cell migration without any chemical factors.

(a)

(b)

(c)

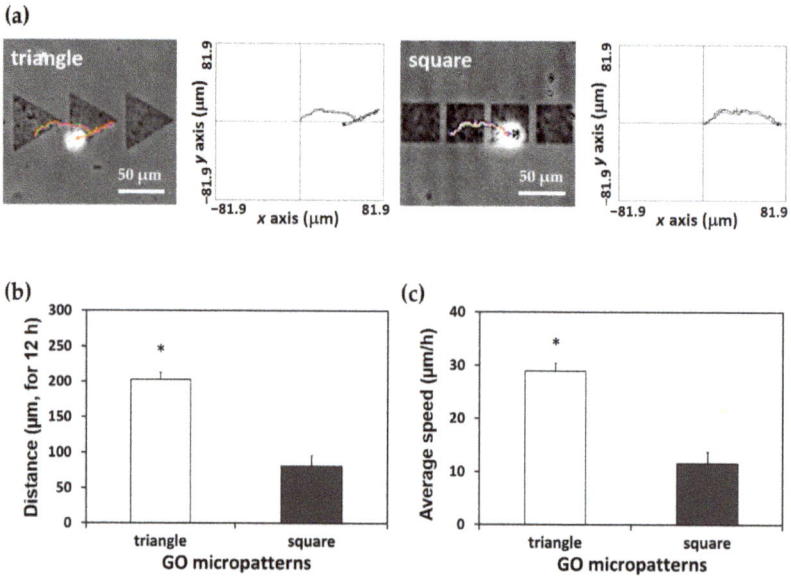

Figure 5. Quantitative analysis of cell migration. (**a**) Trajectories of L-929 fibroblasts on triangle and square GO micropatterns. (**b**) Migration distance and (**c**) average migration speed of L-929 fibroblasts on GO micropatterns. An asterisk (*) denotes a significant difference compared to the square GO micropatterns ($p < 0.05$).

4. Conclusions

The aim of the present study was to develop uniform GO micropatterns and to explore their effects on cell migration. The triangle and square GO micropatterns were finely fabricated using MDD and photolithography techniques. In addition, our findings revealed that the cell migration can be guided by the GO micropatterns having specific geometrical cues, and the triangle GO micropatterns can enhance the cell migration distance, speed, and directionality compared with the square GO micropatterns. Therefore, it is suggested that the GO micropatterns can be employed as a promising biofunctional platform and cell-guiding substrate for applications to bioelectric devices, cell-on-a-chip, and tissue engineering scaffolds.

Supplementary Materials: The following are available online at www.mdpi.com/2072-666X/7/10/186/s1: Video S1: Time-lapse video of L-929 fibroblasts on triangle GO micropatterns for 12 h; Video S2: Time-lapse video of L-929 fibroblasts on square GO micropatterns for 12 h.

Acknowledgments: This work was supported by the Bio and Medical Technology Development Program of the National Research Foundation (NRF) funded by the Korean government (MEST) (No. 2015M3A9E2028643) and the NRF grants funded by the Korean government (MISP) (Nos. 2014R1A2A1A11051704 and 2015R1A5A7036513).

Author Contributions: S.E.K. and M.S.K. designed the experiments, fabricated the GO micropatterns via MDD and photolithography, participated in the cell migration analysis, and drafted the manuscript. Y.C.S. and S.U.E. carried out the characterizations of GO micropatterns. J.H.L. and D.-M.S. prepared the GO and contributed in the cell cultures. S.W.H. and B.K. participated in the quantitative analysis of cell migration. J.-C.P. and B.S.S. performed the statistical analysis and helped interpret the data. D.L. and D.-W.H. conceived of the study, participated in its design and coordination, and helped to draft the manuscript. All authors read and approved the final manuscript.

Conflicts of Interest: The authors declare no conflict of interest.

References

1. Kim, K.H.; Jeong, D.-W.; Jang, N.-S.; Ha, S.-H.; Kim, J.-M. Extremely stretchable conductors based on hierarchically-structured metal nanowire network. *RSC Adv.* **2016**, *6*, 56896–56902. [CrossRef]
2. Voldman, J.; Gray, M.L.; Schmidt, M.A. Microfabrication in biology and medicine. *Annu. Rev. Biomed. Eng.* **1999**, *1*, 401–425. [CrossRef] [PubMed]
3. Kim, M.J.; Lee, J.H.; Shin, Y.C.; Jin, L.; Hong, S.W.; Han, D.-W.; Kim, Y.-J.; Kim, B. Stimulated myogenic differentiation of C2C12 murine myoblasts by using graphene oxide. *J. Korean Phys. Soc.* **2015**, *67*, 1910–1914. [CrossRef]
4. Shin, Y.C.; Lee, J.H.; Kim, M.J.; Hong, S.W.; Kim, B.; Hyun, J.K.; Choi, Y.S.; Park, J.-C.; Han, D.-W. Stimulating effect of graphene oxide on myogenesis of C2C12 myoblasts on RGD peptide-decorated PLGA nanofiber matrices. *J. Biol. Eng.* **2015**, *9*, 22. [CrossRef] [PubMed]
5. Shin, Y.C.; Lee, J.H.; Jin, O.S.; Kang, S.H.; Hong, S.W.; Kim, B.; Park, J.-C.; Han, D.-W. Synergistic effects of reduced graphene oxide and hydroxyapatite on osteogenic differentiation of MC3T3-E1 preosteoblasts. *Carbon* **2015**, *95*, 1051–1060. [CrossRef]
6. Hajjar, Z.; Rashidi, A.M.; Ghozatloo, A. Enhanced thermal conductivities of graphene oxide nanofluids. *Int. Commun. Heat Mass Transf.* **2014**, *57*, 128–131. [CrossRef]
7. Lee, E.J.; Lee, J.H.; Shin, Y.C.; Hwang, D.-G.; Kim, J.S.; Jin, O.S.; Jin, L.; Hong, S.W.; Han, D.-W. Graphene oxide-decorated PLGA/collagen hybrid fiber sheets for application to tissue engineering scaffolds. *Biomater. Res.* **2014**, *18*, 18–24.
8. Shin, Y.C.; Lee, J.H.; Jin, L.; Kim, M.J.; Kim, Y.-J.; Hyun, J.K.; Jung, T.-G.; Hong, S.W.; Han, D.-W. Stimulated myoblast differentiation on graphene oxide-impregnated PLGA-collagen hybrid fibre matrices. *J. Nanobiotechnol.* **2015**, *13*, 21. [CrossRef] [PubMed]
9. Park, K.O.; Lee, J.H.; Park, J.H.; Shin, Y.C.; Huh, J.B.; Bae, J.-H.; Kang, S.H.; Hong, S.W.; Kim, B.; Yang, D.J. Graphene oxide-coated guided bone regeneration membranes with enhanced osteogenesis: Spectroscopic analysis and animal study. *Appl. Spectrosc. Rev.* **2016**, *51*, 540–551. [CrossRef]
10. Lee, J.H.; Lee, Y.; Shin, Y.C.; Kim, M.J.; Park, J.H.; Hong, S.W.; Kim, B.; Oh, J.-W.; Park, K.D.; Han, D.-W. In situ forming gelatin/graphene oxide hydrogels for facilitated C2C12 myoblast differentiation. *Appl. Spectrosc. Rev.* **2016**, *51*, 527–539. [CrossRef]
11. Ko, Y.; Kim, N.H.; Lee, N.R.; Chang, S.T. Meniscus-dragging deposition of single-walled carbon nanotubes for highly uniform, large-area, transparent conductors. *Carbon* **2014**, *77*, 964–972. [CrossRef]
12. Ko, Y.U.; Cho, S.-R.; Choi, K.S.; Park, Y.; Kim, S.T.; Kim, N.H.; Kim, S.Y.; Chang, S.T. Microlitre scale solution processing for controlled, rapid fabrication of chemically derived graphene thin films. *J. Mater. Chem.* **2012**, *22*, 3606–3613. [CrossRef]
13. Kim, N.H.; Kim, B.J.; Ko, Y.; Cho, J.H.; Chang, S.T. Surface energy engineered, high-resolution micropatterning of solution-processed reduced graphene oxide thin films. *Adv. Mater.* **2013**, *25*, 894–898. [CrossRef] [PubMed]
14. Cho, J.; Ko, Y.; Cheon, K.H.; Yun, H.-J.; Lee, H.-K.; Kwon, S.-K.; Kim, Y.-H.; Chang, S.T.; Chung, D.S. Wafer-scale and environmentally-friendly deposition methodology for extremely uniform, high-performance transistor arrays with an ultra-low amount of polymer semiconductors. *J. Mater. Chem. C* **2015**, *3*, 2817–2822. [CrossRef]
15. Gaio, N.; van Meer, B.; Quirós Solano, W.; Bergers, L.; van de Stolpe, A.; Mummery, C.; Sarro, P.M.; Dekker, R. Cytostretch, an organ-on-chip platform. *Micromachines* **2016**, *7*, 120. [CrossRef]
16. Rao, S.; Tata, U.; Lin, V.K.; Chiao, J.-C. The migration of cancer cells in gradually varying chemical gradients and mechanical constraints. *Micromachines* **2014**, *5*, 13–26. [CrossRef]
17. Malinauskas, M.; Rekštytė, S.; Lukoševičius, L.; Butkus, S.; Balčiūnas, E.; Pečiukaitytė, M.; Baltriukienė, D.; Bukelskienė, V.; Butkevičius, A.; Kucevičius, P. 3D microporous scaffolds manufactured via combination of fused filament fabrication and direct laser writing ablation. *Micromachines* **2014**, *5*, 839–858. [CrossRef]
18. Parker, K.K.; Brock, A.L.; Brangwynne, C.; Mannix, R.J.; Wang, N.; Ostuni, E.; Geisse, N.A.; Adams, J.C.; Whitesides, G.M.; Ingber, D.E. Directional control of lamellipodia extension by constraining cell shape and orienting cell tractional forces. *FASEB J.* **2002**, *16*, 1195–1204. [CrossRef] [PubMed]
19. Mahmud, G.; Campbell, C.J.; Bishop, K.J.; Komarova, Y.A.; Chaga, O.; Soh, S.; Huda, S.; Kandere-Grzybowska, K.; Grzybowski, B.A. Directing cell motions on micropatterned ratchets. *Nat. Phys.* **2009**, *5*, 606–612. [CrossRef]

20. Caballero, D.; Comelles, J.; Piel, M.; Voituriez, R.; Riveline, D. Ratchetaxis: Long-range directed cell migration by local cues. *Trends Cell Biol.* **2015**, *25*, 815–827. [CrossRef] [PubMed]
21. Kumar, G.; Ho, C.C.; Co, C.C. Guiding cell migration using one-way micropattern arrays. *Adv. Mater.* **2007**, *19*, 1084–1090. [CrossRef]
22. Kim, M.S.; Lee, M.H.; Kwon, B.-J.; Koo, M.-A.; Seon, G.M.; Lee, J.H.; Han, D.-W.; Park, J.-C. Golgi polarization effects on infiltration of mesenchymal stem cells into electrospun scaffolds by fluid shear stress: Analysis by confocal microscopy and fourier transform infrared spectroscopy. *Appl. Spectrosc. Rev.* **2016**, *51*, 570–581. [CrossRef]
23. Kim, M.S.; Lee, M.H.; Kwon, B.-J.; Koo, M.-A.; Seon, G.M.; Park, J.-C. Enhancement of human mesenchymal stem cell infiltration into the electrospun poly (lactic-co-glycolic acid) scaffold by fluid shear stress. *Biochem. Biophys. Res. Commun.* **2015**, *463*, 137–142. [CrossRef] [PubMed]
24. Kim, M.S.; Lee, M.H.; Kwon, B.-J.; Koo, M.-A.; Seon, G.M.; Park, J.-C. Golgi polarization plays a role in the directional migration of neonatal dermal fibroblasts induced by the direct current electric fields. *Biochem. Biophys. Res. Commun.* **2015**, *460*, 255–260. [CrossRef] [PubMed]
25. Kim, M.S.; Lee, M.H.; Kwon, B.J.; Seo, H.J.; Koo, M.A.; You, K.E.; Kim, D.; Park, J.C. Control of Neonatal Human Dermal Fibroblast Migration on Poly(lactic-co-glycolic acid)-Coated Surfaces by Electrotaxis. *J. Tissue Eng. Regen. Med.* **2015**. Available online: http://onlinelibrary.wiley.com/doi/10.1002/term.1986/abstract; jsessionid=C98BCABAEBE4278D12956CD07DF1A1AD.f01t01 (accessed on 8 October). [CrossRef] [PubMed]
26. Huang, J.; Kim, F.; Tao, A.R.; Connor, S.; Yang, P. Spontaneous formation of nanoparticle stripe patterns through dewetting. *Nat. Mater.* **2005**, *4*, 896–900. [CrossRef] [PubMed]
27. Losego, M.D.; Moh, L.; Arpin, K.A.; Cahill, D.G.; Braun, P.V. Interfacial thermal conductance in spun-cast polymer films and polymer brushes. *Appl. Phys. Lett.* **2010**, *97*, 011908. [CrossRef]
28. Lee, S.C.; Some, S.; Kim, S.W.; Kim, S.J.; Seo, J.; Lee, J.; Lee, T.; Ahn, J.-H.; Choi, H.-J.; Jun, S.C. Efficient direct reduction of graphene oxide by silicon substrate. *Sci. Rep.* **2015**, *5*, 12306. [PubMed]
29. Ji, Z.; Shen, X.; Li, M.; Zhou, H.; Zhu, G.; Chen, K. Synthesis of reduced graphene oxide/CeO$_2$ nanocomposites and their photocatalytic properties. *Nanotechnology* **2013**, *24*, 115603. [CrossRef] [PubMed]
30. Kim, H.J.; Lee, S.-M.; Oh, Y.-S.; Yang, Y.-H.; Lim, Y.S.; Yoon, D.H.; Lee, C.; Kim, J.-Y.; Ruoff, R.S. Unoxidized graphene/alumina nanocomposite: Fracture- and wear-resistance effects of graphene on alumina matrix. *Sci. Rep.* **2014**, *4*, 5176. [CrossRef] [PubMed]
31. Poellmann, M.J.; Harrell, P.A.; King, W.P.; Johnson, A.J.W. Geometric microenvironment directs cell morphology on topographically patterned hydrogel substrates. *Acta Biomater.* **2010**, *6*, 3514–3523. [CrossRef] [PubMed]
32. Yeung, T.; Georges, P.C.; Flanagan, L.A.; Marg, B.; Ortiz, M.; Funaki, M.; Zahir, N.; Ming, W.; Weaver, V.; Janmey, P.A. Effects of substrate stiffness on cell morphology, cytoskeletal structure, and adhesion. *Cell Motil. Cytoskelet.* **2005**, *60*, 24–34. [CrossRef] [PubMed]
33. Liang, C.-C.; Park, A.Y.; Guan, J.-L. In vitro scratch assay: A convenient and inexpensive method for analysis of cell migration in vitro. *Nat. Protoc.* **2007**, *2*, 329–333. [CrossRef] [PubMed]
34. Trepat, X.; Wasserman, M.R.; Angelini, T.E.; Millet, E.; Weitz, D.A.; Butler, J.P.; Fredberg, J.J. Physical forces during collective cell migration. *Nat. Phys.* **2009**, *5*, 426–430. [CrossRef]

micromachines

MDPI

Article

Three-Dimensional Calcium Alginate Hydrogel Assembly via TiOPc-Based Light-Induced Controllable Electrodeposition

Yang Liu [1], Cong Wu [1], Hok Sum Sam Lai [1], Yan Ting Liu [1], Wen Jung Li [1] and Ya Jing Shen [1,2,*]

[1] Department of Mechanical and Biomedical Engineering, City University of Hong Kong, Hong Kong, China; yliu565@cityu.edu.hk (Y.L.); congwuandy@gmail.com (C.W.); samlai5-c@my.cityu.edu.hk (H.S.S.L.); yantinliu2-c@my.cityu.edu.hk (Y.T.L.); wenjli@cityu.edu.hk (W.J.L.)
[2] Shenzhen Research Institute, City University of Hong Kong, Shenzhen 518057, China
* Correspondence: yajishen@cityu.edu.hk; Tel.: +852-3442-8350

Received: 31 March 2017; Accepted: 15 June 2017; Published: 19 June 2017

Abstract: Artificial reconstruction of three-dimensional (3D) hydrogel microstructures would greatly contribute to tissue assembly in vitro, and has been widely applied in tissue engineering and drug screening. Recent technological advances in the assembly of functional hydrogel microstructures such as microfluidic, 3D bioprinting, and micromold-based 3D hydrogel fabrication methods have enabled the formation of 3D tissue constructs. However, they still lack flexibility and high efficiency, which restrict their application in 3D tissue constructs. Alternatively, we report a feasible method for the fabrication and reconstruction of customized 3D hydrogel blocks. Arbitrary hydrogel microstructures were fabricated in situ via flexible and rapid light-addressable electrodeposition. To demonstrate the versatility of this method, the higher-order assembly of 3D hydrogel blocks was investigated using a constant direct current (DC) voltage (6 V) applied between two electrodes for 20–120 s. In addition to the plane-based two-dimensional (2D) assembly, hierarchical structures—including multi-layer 3D hydrogel structures and vessel-shaped structures—could be assembled using the proposed method. Overall, we developed a platform that enables researchers to construct complex 3D hydrogel microstructures efficiently and simply, which has the potential to facilitate research on drug screening and 3D tissue constructs.

Keywords: three-dimensional (3D) hydrogel assembly; TiOPc; alginate hydrogel; light-induced electrodeposition

1. Introduction

The construction of a cell-friendly three-dimensional (3D) extracellular matrix (ECM) is a major challenge in the fields of biomedical and tissue engineering [1]. The 3D hydrogel-based tissue constructs, such as cellular microarrays and engineered tissue analogues, have been introduced as an alternative to animal experiments for advanced biomedical studies in vitro, pharmacological assays, observation of dynamic cellular process, and tissue morphogenesis using smaller sample volumes [2–4]. In order to generate geometry-controllable 3D hydrogel constructs, microfluidic technologies and 3D printing tools are widely applied. Currently, a variety of hydrogel shapes of particles, fibers, and sheets are used to reconstruct complex 3D tissue scaffolds in a bottom-up approach [5–11].

Artificial reconstruction is a very important strategy to achieve a higher-order assembly of several functional 3D tissue constructs. To implement this strategy, a number of emerging methods related to 3D hydrogel construction are available [12–14]. The micromold-based 3D hydrogel fabrication has high reproducibility due to the simple fabrication process. A number of 3D hydrogel constructs are easily formed in a 3D mold for use in high-throughput drug screening. Nevertheless, one substantial drawback of this method is a lack of flexibility of mold structure [15]. The microfluidic and 3D bioprinting technologies are efficient ways to fabricate 3D hydrogel constructs without the use of a mold. By using a combination of microfluidic and 3D bioprinting technologies, hydrogel blocks, capsules, and microfibers can be fabricated rapidly [5–9,11,16]. However, after the basic hydrogel blocks are constructed, it may be more difficult to reconstruct them into complex 3D tissues or multi-tissue architectures (e.g., blood vessels-shaped).

Recently, a shape control technique of 3D hydrogel construction based on electrodeposition was reported [17–19]. Because of its good biocompatibility and in situ cross-link with the calcium ions (Ca^{2+}), Ca-alginate hydrogel has been widely applied to entrap and immobilize cells for the construction of 3D cellular tissue [3,20–22]. By means of electrodeposition, the 3D hydrogel patterns are generated on a 2D microelectrode surface, based on which in-situ 3D gel structures are fabricated with controllable size and shape. It is promising to fabricate the 3D gel structures rapidly independent of the 3D mold. Nevertheless, the electrode patterns still require pre-fabrication via mask microfabrication methods, such as photolithography and wet etching techniques [23]. Once a given geometry is fabricated, the resulting pattern is fixed as well. Hence, a more flexible and simple 3D gel fabrication strategy is essential for efficient 3D tissue reconstruction.

In this paper, we report an easy-to-use and universal approach for the flexible and rapid fabrication of 3D hydrogel blocks of Ca-alginate and readily assemble them into multi-layer or planar welding tissue constructs. In contrast to our previous efforts [12–15], a mold-free light-addressable electrodeposition method was adopted for this work. The method enables the formation of complex 3D hydrogel constructs via organic photoconduction-based controllable 3D hydrogel patterning. Herein, a photoconductive chip was developed based on titanium oxide phthalocyanine (TiOPc) due to its broad absorption from visible to infrared regions [24]. The TiOPc layer allows the impedance to be tuned by illumination. Therefore, coating TiOPc on indium tin oxide (ITO) glass yields a virtual electrode. Consequently, the electrode pattern is controlled via a visible-light projection onto the TiOPc chip. In this work, a Ca-alginate solution—a biocompatible hydrogel—was used to format the 3D tissue constructs. A programmable 3D hydrogel micropattern was fabricated using custom-designed optical patterns and a direct current (DC) electric field, which enabled Ca^{2+} cross-linking with alginate to form a 3D hydrogel microstructure in situ. The method provides a more effective fabrication platform for the 3D tissue constructs' assembly into multi-layer or planar welding 3D bioconstructs, which has the potential to promote research into cell interaction mechanisms, single cell culture, and tissue reconstruction.

2. Materials and Methods

2.1. Chip Design and Fabrication

First, 500 µL TiOPc solution was dropped onto the top surface of a 30 mm × 30 mm indium tin oxide (ITO) glass, which was spun at a speed 500 rpm for 15 s and accelerated at a speed 1200 rpm for 60 s, resulting in a layer of approximately 10 µm. Next, the coated plate was baked at 120 °C for 30 min to harden the TiOPc layer. The fabrication process is illustrated in Figure 1c.

The photoconductive chip was composed of three parts: a top ITO glass served as one of the electrodes, a bottom ITO glass surface coated with a thin layer of TiOPc as a light-addressable electrode, and a Ca-alginate hydrogel as an electrically-induced deposition solution. It represented a sandwich structure, as shown in Figure 1b. The distance between the two electrodes was adjusted according to the height of the fabricated 3D hydrogel structures from 40 µm to 1500 µm.

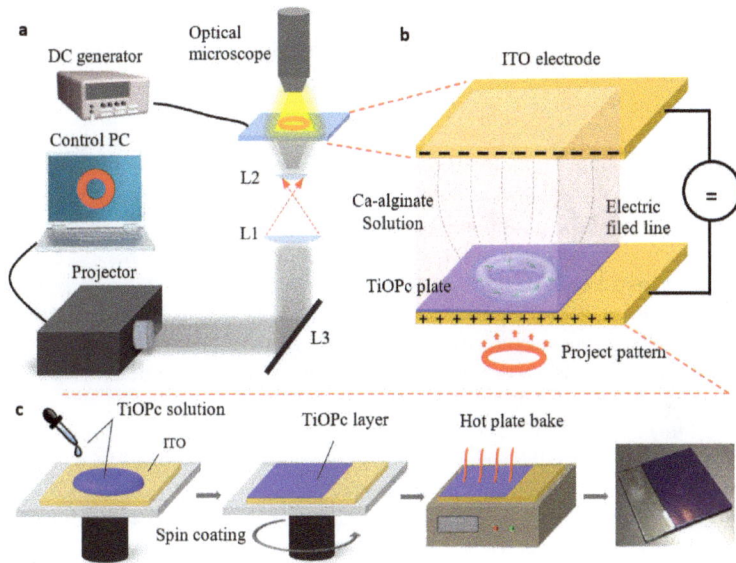

Figure 1. Schematic of the rapid light-addressable system and preparation process of the titanium oxide phthalocyanine (TiOPc) plate. (**a**) The programmable light patterns are projected onto the photoconductive chip through contraction lenses L1 and L2. The 3D hydrogel fabrication processes was observed under an optical microscope with a 10× objective lens and a charge-coupled device (CCD) camera. (**b**) The top indium tin oxide (ITO) glass and bottom TiOPc plate represent the two electrodes of the photoconductive chip. During the electrodeposition process, a Ca-alginate solution was introduced between the two electrodes. A DC electric field and an optical pattern triggered Ca-alginate solution cross-linking to generate 3D hydrogel constructs based on the virtual electrode pattern. (**c**) A TiOPc layer was generated by spin-coating at a certain spinning speed. Its thickness can be modified by adjusting the spinning speed.

2.2. Preparation of Deposition Solution

Sodium alginate powder (Sigma Inc., Marlborough, MA, USA) was dissolved in distilled water at 1% (w/v). Then, insoluble $CaCO_3$ powder (diameter 30–50 nm; Haofu Chemistry Limited Company, Shanghai, China) was dispersed into the solution at 0.5% (w/v) followed by magnetic stirring at 1200 rpm for 12 h.

2.3. Fluorescent Imaging

To obtain the fluorescent images of 3D hydrogel structures, 1% (w/v) suspension of the fluorescent microspheres (10 μm in diameter, Aladdin Inc., Shanghai, China) was mixed with the deposition solution and sonicated for 10 min. A fluorescence microscope (Eclipse Ni, Nikon, Tokyo, Japan) was used to observe the stained hydrogel at 4× magnification.

2.4. Formation of the 3D Hydrogel Microcapsule

To prevent the breaking of the hydrogel during reconstruction, the alginate-PLL (Poly-L-Lysine)microcapsule was fabricated as a shell. After the fabrication of the 3D hydrogel structures, the TiOPc plate was immersed into a 10 cm petri dish to flush away extra deposition solution. Then, the 3D hydrogel structures were detached from the TiOPc plate surface and transferred into another petri dish filled with 0.05% (w/v) PLL solution by gentle pipetting. The Ca-alginate hydrogel structures reacted with the PLL solution for 5 min, and the alginate–PLL membrane was formed. Next, 1.1%

(w/v) CaCl$_2$ solution was used to harden the alginate–PLL gel structures. The hardened 3D hydrogel structures were then treated with 0.03% (w/v) sodium alginate solution for 4 min to obtain the final microcapsule shell.

3. Results

For electrodeposition of Ca-alginate hydrogel, a rapid light-addressable system was used as shown in Figure 1a. The illumination source was provided by a projector. Then, the optical patterns programmed via a commercial computer were projected onto the TiOPc light-addressable electrode. The optical image contraction system consisted of two lenses of focal lengths (L1 and L2) 200 mm and 20 mm, respectively, and a dichroic mirror (L3), which served to minimize the optical pattern 1/10×. A DC potential applied between two electrodes was supplied by a signal generator (2601B, Keithley, Cleveland, OH, USA). The photoconductive chip was placed on a 3D translation platform, and the 3D hydrogel microstructures fabrication processes were monitored under an optical microscope with 10× objective lens and a charge-coupled device (CCD) camera.

The procedure of optically-induced 3D Ca-alginate hydrogel electrodeposition is illustrated in Figure 2. As mentioned earlier, the impedance of TiOPc could be tuned by illumination. Hence, the TiOPc layer was insulated when no light was projected on it, even when the DC source was applied between the two electrodes. As a consequence, electrically-induced deposition could not be triggered, as shown in Figure 2a. Whereas, after the optical pattern was projected on the TiOPc electrode, the impedance was adjusted by the projected area and the TiOPc electrode was conductive immediately. The photoconductive chip formed into a circuit, and the Ca-alginate hydrogel electrodeposition triggered on the positive electrode. Herein, a DC voltage of constant potential (6 V) was applied between the two electrodes for 20–120 s. The electric field across the solution generated H$^+$ by electrolysis of water and formed a pH gradient around the anode surface (TiOPc layer), as shown in Figure 2b.

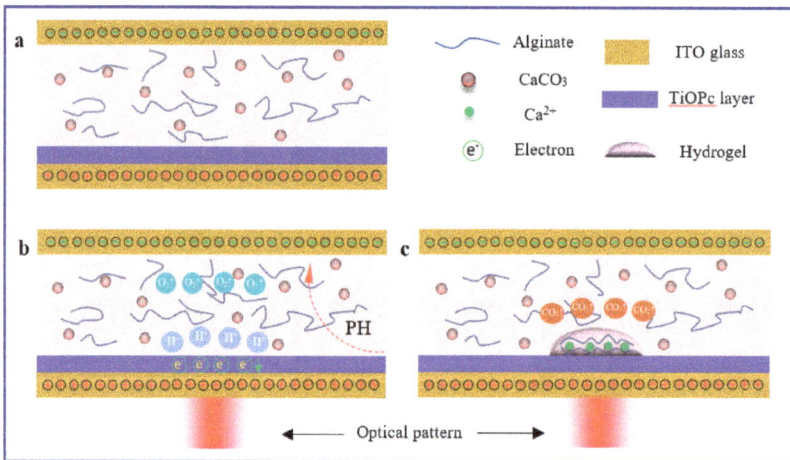

Figure 2. Schematic of light-induced electrodeposition. (**a**) The Ca-alginate solution was introduced between the cathodic ITO glass and the anodic TiOPc plate. (**b**) Light and DC voltage triggered the production of H$^+$ nearby the anode, leading to a decreased pH gradient. (**c**) H$^+$ reacted with CaCO$_3$ nanoparticles in the solution to release Ca^{2+}. After that, alginate cross-linking with the release of Ca^{2+} formed the 3D hydrogel structures based on the virtual electrode patterns.

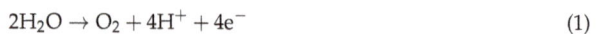

$$2H_2O \rightarrow O_2 + 4H^+ + 4e^- \tag{1}$$

After that, the nano $CaCO_3$ particles in the solution reacted with H^+ to release calcium ions (Ca^{2+}).

$$2H^+ + CaCO_3 \rightarrow H_2O + Ca^{2+} + CO_2 \tag{2}$$

Finally, alginate in the solution cross-linked with the released Ca^{2+} to form the 3D Ca-alginate hydrogel structure onto the TiOPc plate corresponding to the virtual electrode pattern (Figure 2c).

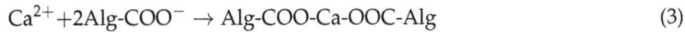

$$Ca^{2+} + 2Alg\text{-}COO^- \rightarrow Alg\text{-}COO\text{-}Ca\text{-}OOC\text{-}Alg \tag{3}$$

During the growth of the Ca-alginate hydrogel by electrodeposition, the density of Ca^{2+} was increased near the anode surface during the electrodeposition process as the time increased. Likewise, the gel's height and hardness increased with time, and eventually reached a steady state. Thus, the formation of the Ca-alginate hydrogel is related to current density and period of illumination. The relationship between them has been exhibited in Figure 3. The dependence of deposition time on the height of gel growth is illustrated in Figure 3a. The trend of the gel growth increased linearly in time and finally reached approximately 400 μm at the constant current density of 3 Am^{-2}. The effect of current density on hydrogel growth at the constant deposition time of 90 s was measured, which is shown in Figure 3b. The fit curve clearly confirms the relationship between the current density and the height of gel growth; i.e., the hydrogel grew faster under a higher current density.

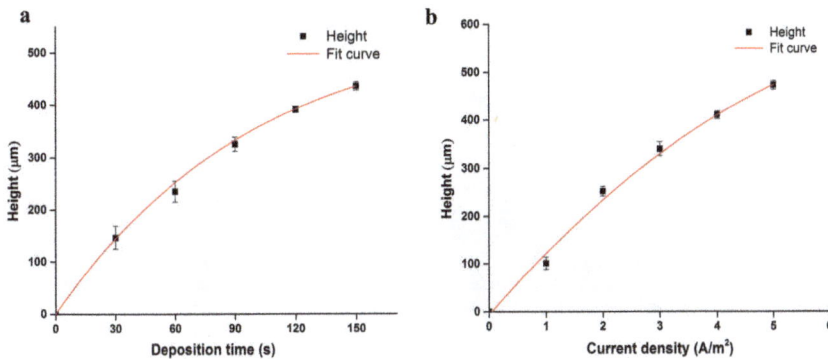

Figure 3. (a) The dependence of the deposition time on the height of the gel growth at constant current density 3 Am^{-2}. The root mean squared error (RMSE) between experiment and model is 5.9 μm. (b) The effect of the current density on the hydrogel growth. The RMSE between experiment and model is 0.1 μm.

To demonstrate the capability of the 3D hydrogel microstructures assembly achieved by this method, the three main types of 3D hydrogel construction assembly strategies were investigated systematically. The plane-based two-dimensional (2D) assembly is the common artificial construction strategy. The process of reconstructing two squared 3D hydrogel blocks into a full rectangular 3D hydrogel structure by planar welding is depicted in Figure 4. Red and green fluorescent microspheres were mixed into the Ca-alginate solution respectively for distinguishing the different 3D hydrogel blocks. A square optical pattern was projected on the TiOPc plate, and a red, squared 3D hydrogel block was fabricated via the light-addressable electrodeposition firstly, as shown in Figure 4a. After 50 s, the TiOPc plate was immersed into a petri dish (50 cm diameter) containing deionized water to gently flush away the extra alginate, and the boundary of the squared 3D hydrogel structure was defined clearly (Figure 4b). Again, a Ca-alginate solution mixed with green fluorescent microspheres was dropped on the TiOPc plate and joint with the red-square 3D hydrogel block. A squared optical pattern was projected on the TiOPc plate close to or slightly overlying the area of the red-squared 3D hydrogel block (Figure 4c). The same 3D hydrogel block fabrication process was carried out for the green-squared

3D hydrogel block, and a full rectangular 3D hydrogel structure was obtained via planar welding of two 3D squared hydrogel blocks (Figure 4d). Figure 4e–g show the assembly of the two 3D squared hydrogel blocks successfully. To avoid breaking the fragile hydrogels, an alginate–PLL membrane was formed as a shell to wrap the 3D hydrogel structure, sequentially forming a hydrogel microcapsule.

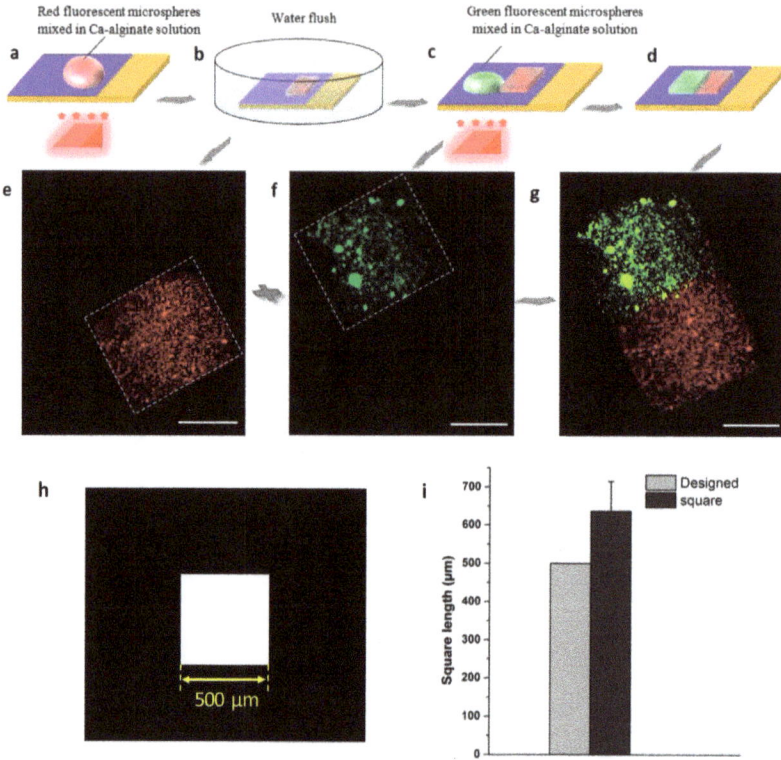

Figure 4. Reconstruction of 3D hydrogel blocks via planar welding. (**a**) The 3D hydrogel microstructure (red-square) was fabricated via a light-addressable electrodeposition system. (**b**) The 3D hydrogel structure came out after the deionized water gently flushed away the extra solution. (**c**) The Ca-alginate solution mixed with green fluorescent microspheres was dropped on the TiOPc layer adjacent next to the red-square 3D hydrogel structure, and the same hydrogel fabrication procedure was applied on it again. (**d**) After 50 s, the full-rectangle 3D hydrogel structure was achieved via planar welding of the green-square 3D hydrogel block with the red one. (**e–g**) The fluorescence images of the planar-weld 3D hydrogel structure. All scale bars are 250 μm. (**h**) The designed patterns of the virtual electrode. (**i**) Size distribution of designed pattern and 3D hydrogel microstructure.

The photoconductive electrodes allowed the Ca-alginate hydrogel to form into a desired pattern at a specific address. To demonstrate the flexibility and reliability of this 3D gel microstructure fabrication strategy, various 3D hydrogel blocks were fabricated according to the virtual electrode patterns at will and were then assembled into the desired 3D tissue constructions in the condition of constant potential voltage 6 V and deposition time 20–120 s. The method of reconstructing 3D hydrogel blocks into new 3D hydrogel architecture was developed by the light-addressable electrodeposition. Consequently, it is suggested that the present method will allow researchers to fabricate more complex 3D tissues in vitro (e.g., blood vessels, muscle fibers, and neural pathways). It is noteworthy that they are the hierarchical structures of the human body.

Besides the planar welding, the hierarchical structures can be also assembled by the flexible and rapidly light-addressable electrodeposition method, which are useful for the creation of various complex functional objects, such as multi-layer 3D hydrogel structures and vessel-shaped structures. Figure 5 shows the building process of the multi-layer 3D hydrogel structures. A red-hexagonal 3D hydrogel structure was fabricated via light-addressable electrodeposition as the first layer, as shown in Figure 5a,b, and a 3D hydrogel structure based on a desired optical pattern was obtained, as shown in Figure 5e. For the second layer, green fluorescent microspheres were mixed with 1% (w/v) sodium alginate solution and dropped on the anode to overlay the area of the first layer, which serves as the feeder layer for the provision of free Ca^{2+}. The first feeder layer contacts with the sodium alginate solution, then the anode current and optical pattern together trigger the Ca^{2+}, which is released from the excess $CaCO_3$ nanoparticles in the feeder layer to form the second layer of the Ca-alginate cross-linking hydrogel structures (Figure 5c,d). Herein, a circular optical pattern was projected on the overlying area as the second layer. After 60 s, a green-circular 3D hydrogel structure was generated on the first layer, and later the multi-layered structure was assembled successfully. The experimental results are shown in Figure 5f,g, respectively.

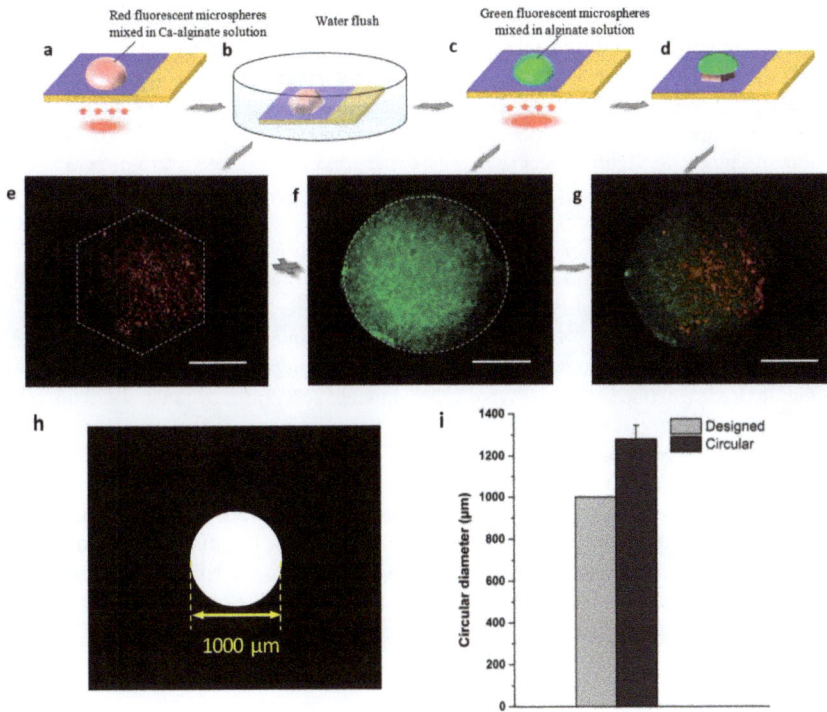

Figure 5. Formation of multi-layer 3D hydrogel microstructure. (**a**) A hexagonal optical pattern was projected on the TiOPc plate, and a red 3D hydrogel structure formed in the anode as the first layer. (**b**) After deposition, the deionized water was used to flush away the excess hydrogel solution. (**c**) Based on the first layer, the hydrogel solution without $CaCO_3$ nanoparticles was dropped on it, and a circular optical pattern was projected on the overlying area to form a second layer. (**d**) A multi-layer 3D hydrogel microstructure assembled with the red hexagonal 3D hydrogel block and the green circular 3D hydrogel block. (**e–g**) The fluorescence images of the multi-layer 3D hydrogel structure. All scale bars are 250 μm. (**h**) The designed patterns of the virtual electrode. (**i**) Size distribution of designed pattern and 3D hydrogel microstructure.

The assembly process of the vessel-shaped structure is shown in Figure 6. To achieve a long and thin vessel-shaped 3D hydrogel structure, the optical pattern was projected on the TiOPc plate and moved along the Ca-alginate solution to cover the area step by step, with a deposition time of 50 s in each step. The consecutive electrodeposition was achieved as the designed length, the inner layer of the vessel-shaped structure was obtained as shown in Figure 6a. The fabricated inner layer can be clearly present via the deionized water flush away the extra Ca-alginate solution (Figure 6b). Herein, a squared optical pattern was used as the electrodeposition unit, and the length of the inner layer was approximately 2000 μm, while its width was approximately 400 μm. The fluorescence images of the vessel-shaped hydrogel structure are shown in Figure 6e. The fabrication process of the outer layer of the vessel-shaped structure is shown in Figure 6c,d. After flushing away the excess deposition solution, the inner layer 3D hydrogel structure was detached from the TiOPc plate by gentle pipetting and transferred into another 10 cm petri dish filled with the sodium alginate solution mixed green fluorescent microspheres. After treatment with the sodium alginate solution for 5 min, the fabricated inner layer was wrapped sufficiently. The wrapped inner layer was transferred to the TiOPc plate by pipetting gently. The same procedure as followed for the inner layer was used for the fabrication of the outer layer. After 50 s, an outer layer was fabricated with a length of approximately 2000 μm and width of approximately 450 μm. The results of the experiment are shown in Figure 6f. Figure 6g shows the successful assembly of the 3D vessel-shaped hydrogel microstructure.

For this study, we fabricated various 3D hydrogel microstructures using this method. Figure 3 shows two square-shaped 3D hydrogel microstructures with length approximately 500 μm, while the length of the designed micro-pattern was 450 μm. The results indicate that the 3D hydrogel microstructures might be slightly larger than their corresponding designed micro-pattern because of the diffusion of calcium ions during the electrodeposition processes. We investigated the relationship between the designed optical pattern and the 3D hydrogel microstructures quantitatively. Three different micro-patterns were designed—squared with length 500 μm (Figure 4h), circular with diameter 1000 μm (Figure 5h), and rectangular with length 2000 μm and width 400 μm (Figure 6h), respectively. After light-induced electrodeposition, the length of the 2D square hydrogel microstructure was approximately 637 μm ± 78 μm (Figure 4i). The diameter of the 2D circular hydrogel microstructure was approxim0ately 1279 μm ± 67 μm (Figure 5i). The length and width of the 2D rectangular hydrogel microstructure were approximately 2430 μm ± 39 μm and 479 μm ± 45 μm (Figure 6i), respectively.

Consequently, the proposed method realized various 3D hydrogel blocks and the highly-efficient reconstruction of complex 3D microstructures. The advantage of this method is the capability to fabricate both plane-based 2D assembly and spatial assembly. Moreover, the structure and the shape of the 3D hydrogel constructs can be fabricated controllably (i.e., the length, width, thickness, and the layer numbers of hierarchical structures). The high flexibility and controllability of this method could have the potential to improve fabrication efficiency and the application of 3D hydrogel architectures.

Micromachines **2017**, *8*, 192

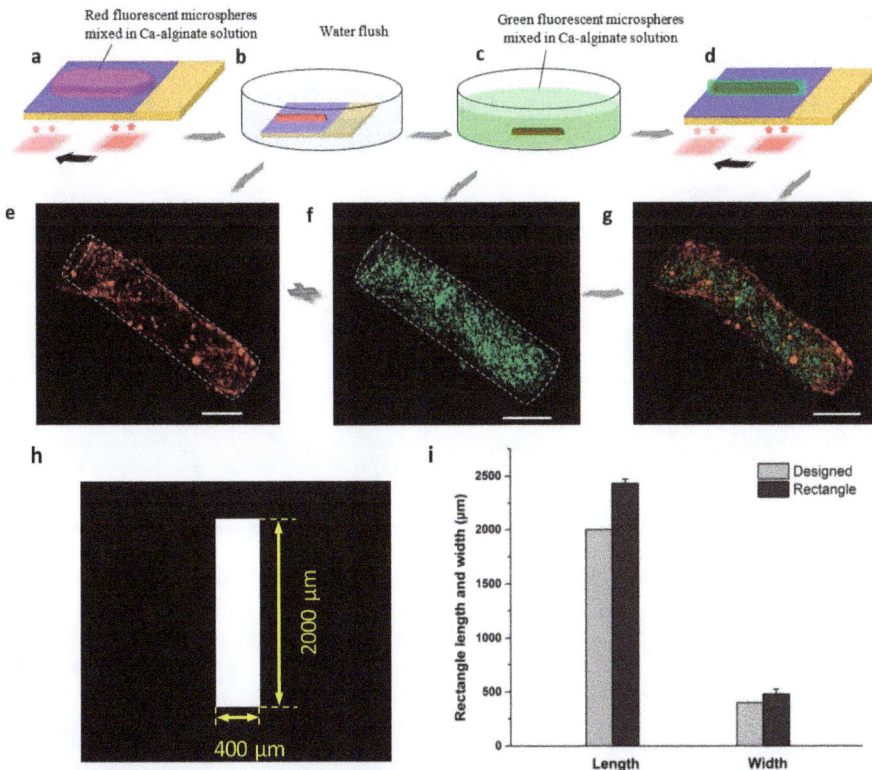

Figure 6. Assembly processes of the vessel-shaped structure. (**a**) The inner layer was fabricated via light-addressable electrodeposition step by step via optical pattern moved along the deposition solution cover area. (**b**) The TiOPc plate was immersed into the deionized water to flush away the extra deposition solution. (**c**) The formed inner layer was detached by pipetting gently and transferred into the sodium alginate solution mixed with green fluorescent microspheres. (**d**) The same fabrication steps were used to form the outer layer. (**e–g**) The fluorescence images of the vessel-shaped hydrogel structure. All scale bars are 400 μm. (**h**) The designed patterns of the virtual electrode. (**i**) Size distribution of designed pattern and 3D hydrogel microstructure.

4. Conclusions

In summary, we described a system for the study of the highly-efficient fabrication and flexible reconstruction of 3D hydrogel microstructures. The 3D hydrogel blocks can be fabricated readily and rapidly via mold-free light-addressable electrodeposition. The simple fabrication method presented herein allows the generation of desired hydrogel structures at a specific address. Furthermore, the method requiring only general electrodeposition and hydrogel fabrication processes can produce the construction of complex 3D hydrogel architectures without sophisticated techniques. Conclusively, the versa tile method holds metapromise as a universal platform for artificial and functional architectures applicable to drug screening and tissue transplantation.

Supplementary Materials: The following are available online at www.mdpi.com/2072-666X/8/6/192/s1, Figure S1: The confocal fluorescence image of the 3D hydrogel microstructure. The scale bar is 200 μm.

Acknowledgments: This work was support by National Science Foundation of China (61403323), RGC General Research Fund of Hong Kong (CityU 11217915), and ShenZhen (China) Basic Research Project (JCYJ20160329150236426).

Author Contributions: Yang Liu designed and performed the experiments and wrote the paper; Cong Wu, Hok Sum Sam Lai, and Yan Ting Liu contributed reagents and analysis tools; Wen Jung Li and Ya Jing Shen commented on the manuscript.

Conflicts of Interest: The authors declare no conflict of interest.

References

1. Fernandes, T.G.; Diogo, M.M.; Clark, D.S.; Dordick, J.S.; Cabral, J.M.S. High-throughput cellular microarray platforms: Applications in drug discovery, toxicology and stem cell research. *Trends Biotechnol.* **2009**, *27*, 342–349. [CrossRef] [PubMed]

2. Nie, Z.; Kumacheva, E. Patterning surfaces with functional polymers. *Nat. Mater.* **2008**, *7*, 277–290. [CrossRef] [PubMed]

3. Tan, W.H.; Takeuchi, S. Monodisperse alginate hydrogel microbeads for cell encapsulation. *Adv. Mater.* **2007**, *19*, 2696–2701. [CrossRef]

4. Lee, K.H.; Shin, S.J.; Park, Y.; Lee, S.H. Synthesis of cell-laden alginate hollow fibers using microfluidic chips and microvascularized tissue-engineering applications. *Small* **2009**, *5*, 1264–1268. [CrossRef] [PubMed]

5. Onoe, H.; Okitsu, T.; Itou, A.; Takeuchi, S. Metre-long cell-laden microfibres exhibit tissue morphologies and functions. *Nat. Mater.* **2013**, *12*, 584–590. [CrossRef] [PubMed]

6. Yamada, M.; Utoh, R.; Ohashi, K.; Tatsumi, K.; Yamato, M.; Okano, T.; Seki, M. Controlled formation of heterotypic hepatic micro-organoids in anisotropic hydrogel microfibers for long-term preservation of liver-specific functions. *Biomaterials* **2012**, *33*, 8304–8315. [CrossRef] [PubMed]

7. Yamada, M.; Sugaya, S.; Naganuma, Y.; Seki, M. Microfluidic synthesis of chemically and physically anisotropic hydrogel microfibers for guided cell growth and networking. *Soft Matter* **2012**, *8*, 3122–3130. [CrossRef]

8. Lee, D.H.; Lee, W.; Um, E.; Park, J.K. Microbridge structures for uniform interval control of flowing droplets in microfluidic networks. *Biomicrofluidics* **2011**, *5*, 034117. [CrossRef] [PubMed]

9. Um, E.; Lee, D.S.; Pyo, H.B.; Park, J.K. Continuous generation of hydrogel beads and encapsulation of biological materials using a microfluidic droplet-merging channel. *Microfluid. Nanofluidics* **2008**, *5*, 541–549. [CrossRef]

10. Lee, W.; Bae, C.Y.; Kwon, S.; Son, J.; Kim, J.; Jeong, Y.; Yoo, S.S.; Park, J.K. Cellular hydrogel biopaper for patterned 3D cell culture and modular tissue reconstruction. *Adv. Healthc. Mater.* **2012**, *1*, 635–639. [CrossRef] [PubMed]

11. Shembekar, N.; Chaipan, C.; Utharala, R.; Merten, C.A. Droplet-based microfluidics in drug discovery, transcriptomics and high-throughput molecular genetics. *Lab Chip* **2016**, *16*, 1314–1331. [CrossRef] [PubMed]

12. Khademhosseini, A.; Ferreira, L.; Yeh, J.; Karp, J.M.; Fukuda, J.; Langer, R. Co-culture of human embryonic stem cells with murine embryonic fibroblasts on microwell-patterned substrates. *Biomaterials* **2006**, *27*, 5968–5977. [CrossRef] [PubMed]

13. Karp, J.M.; Yeh, J.; Eng, G.; Fukuda, J.; Blumling, J.; Suh, K.Y.; Cheng, J.; Mahdvai, A.; Borenstein, J.; Langer, R.; et al. Controlling size, shape and homogeneity of embryoid bodies using poly (ethylene glycol) microwells. *Lab Chip* **2007**, *7*, 786–794. [CrossRef] [PubMed]

14. Moeller, H.C.; Mian, M.K.; Shrivastava, S.; Chung, B.G.; Khademhosseini, A. A microwell array system for stem cell culture. *Biomaterials* **2008**, *29*, 752–763. [CrossRef] [PubMed]

15. Matsunaga, Y.T.; Morimoto, Y.; Takeuchi, S. Molding cell beads for rapid construction of macroscopic 3D tissue architecture. *Adv. Mater.* **2011**, *23*, H90–H94. [CrossRef] [PubMed]

16. Tan, Y.; Richards, D.J.; Trusk, T.C.; Visconti, R.P.; Yost, M.J.; Kindy, M.S. 3D printing facilitated scaffold-free tissue unit fabrication. *Biofabrication* **2014**, *6*, 024111. [CrossRef] [PubMed]

17. Betz, J.F.; Cheng, Y.; Tsao, C.Y.; Zargar, A.; Wu, H.C.; Luo, X.; Payne, G.F.; Bentley, W.E.; Rubloff, G.W. Optically clear alginate hydrogels for spatially controlled cell entrapment and culture at microfluidic electrode surfaces. *Lab Chip* **2013**, *13*, 1854–1858. [CrossRef] [PubMed]

18. Shi, X.W.; Tsao, C.Y.; Yang, X.; Liu, Y.; Dykstra, P.; Rubloff, G.W.; Ghodssi, R.; Bentley, W.E.; Payne, G.F. Electroaddressing of cell populations by co-deposition with calcium alginate hydrogels. *Adv. Funct. Mater.* **2009**, *19*, 2074–2080. [CrossRef]

19. Cheng, Y.; Luo, X.; Tsao, C.Y.; Wu, H.C.; Betz, J.; Payne, G.F.; Bentley, W.E.; Rubloff, G.W. Biocompatible multi-address 3D cell assembly in microfluidic devices using spatially programmable gel formation. *Lab Chip* **2011**, *11*, 2316–2318. [CrossRef] [PubMed]
20. Derby, B. Printing and prototyping of tissues and scaffolds. *Science* **2012**, *338*, 921–926. [CrossRef] [PubMed]
21. Capretto, L.; Mazzitelli, S.; Balestra, C.; Tosi, A.; Nastruzzi, C. Effect of the gelation process on the production of alginate microbeads by microfluidic chip technolog. *Lab Chip* **2008**, *8*, 617–621. [CrossRef] [PubMed]
22. Dai, G.; Wan, W.; Zhao, Y.; Wang, Z.; Li, W.; Shi, P.; Shen, Y. Controllable 3D alginate hydrogel patterning via visible-light induced electrodeposition. *Biofabrication* **2016**, *8*, 025004. [CrossRef] [PubMed]
23. Kang, Y.; Cetin, B.; Wu, Z.; Li, D. Continuous particle separation with localized AC-dielectrophoresis using embedded electrodes and an insulating hurdle. *Electrochimica Acta* **2009**, *54*, 1715–1720. [CrossRef]
24. Wang, W.B.; Li, X.G.; Wang, S.R.; Hou, W. The preparation of high photosensitive TiOPc. *Dyes Pigment.* **2007**, *72*, 38–41. [CrossRef]

micromachines

MDPI

Review

Microfluidic Technology for the Generation of Cell Spheroids and Their Applications

Raja K. Vadivelu [1], Harshad Kamble [2], Muhammad J. A. Shiddiky [1,2] and Nam-Trung Nguyen [2,*]

[1] School of Natural Sciences, Nathan Campus, Griffith University, 170 Kessels Road, Brisbane, QLD 4111, Australia; raja.vadivelu@griffithuni.edu.au (R.K.V.); m.shiddiky@griffith.edu.au (M.J.A.S.)
[2] QLD Micro- and Nanotechnology Centre, Nathan Campus, Griffith University, 170 Kessels Road, Brisbane, QLD 4111, Australia; harshad.kamble@griffithuni.edu.au
* Correspondence: nam-trung.nguyen@griffith.edu.au; Tel.: +61-07-373-53921

Academic Editors: Aaron T. Ohta and Wenqi Hu
Received: 11 January 2017; Accepted: 15 March 2017; Published: 23 March 2017

Abstract: A three-dimensional (3D) tissue model has significant advantages over the conventional two-dimensional (2D) model. A 3D model mimics the relevant in-vivo physiological conditions, allowing a cell culture to serve as an effective tool for drug discovery, tissue engineering, and the investigation of disease pathology. The present reviews highlight the recent advances and the development of microfluidics based methods for the generation of cell spheroids. The paper emphasizes on the application of microfluidic technology for tissue engineering including the formation of multicellular spheroids (MCS). Further, the paper discusses the recent technical advances in the integration of microfluidic devices for MCS-based high-throughput drug screening. The review compares the various microfluidic techniques and finally provides a perspective for the future opportunities in this research area.

Keywords: microfluidics; bioMEMS; cell spheroids; three-dimensional cell culture; tissue engineering

1. Introduction

In the past decade, cell-based assays have undergone a noticeable transition from two-dimensional (2D) to three-dimensional (3D) cell culture. A cell culture is the basic tool for drug discovery, investigation of the mechanism of diseases, and tissue engineering. A 3D cell culture maintains the significant physiological relevance of cell-based assays [1]. A 3D cell culture mimics the sophisticated in-vivo environment which is crucial for efficiently predicting the mechanisms of drug action before clinical trials. Traditionally, 2D cell cultures on a flat substrate are employed as in-vitro models, because they are inexpensive and more accessible than animal models. However, 2D culture models may not be able to mimic the in-vivo systems in terms of cellular physiology, metabolism and protein expression (e.g., membrane proteins). Current literature indicates that the spatially confined 2D cultures attribute to the forced inhabitation of cells grown on a flat and rigid surface [2]. The flat surface requires cytoskeleton to establish contact between neighbouring cells and exert artificial polarity [3]. Thus, 2D cultures cannot provide adequate extracellular matrix (ECM) formation and promote cell–cell and cell–matrix interaction to form a complex communication network within a tissue-specific architecture [4]. ECM is a critical cellular factor for structural support and biochemical cues that regulate cell proliferation, adhesion and migration. Furthermore, cells in a monolayer are exposed to the bulk of media with sufficient oxygen and nutrients, whereas the response of cells in a 3D tissue to nutrient and soluble factors depends on their diffusion and the corresponding concentration distribution [5,6].

The limitations of 2D culture systems motivate the development of 3D culture. In contrast to the flat 2D culture, a 3D culture consists of multi-cellular layers, which are critical for both biochemical and mechanical characteristics of a tissue. Thus, a 3D construct allows for the optimal transport of nutrient, gas, growth factors and cellular waste similar to in-vivo processes. To date, countless efforts have been reported on the production of more biologically relevant 3D tissue models using both scaffold-based and scaffold-free strategies. Microtissues constructed with scaffold rely on supporting materials, which raises issues of biocompatibility and cell–material biorecognition. Biodegradable scaffold substitutes a large amount ECM, resulting in tissue that is composed of less densely packed cells [7]. Furthermore, biodegradable scaffolds exert sensitivity to standard sterilization method when used as an implant in the surgical site. In contrast, scaffold-free approaches initiate interactions between cells and substrate to maximize cell–cell interaction by self-generated ECM. In recent years, scaffold-free methods have been developed to enable the self-assembly of cells into multi-planar cell sheets or spherical cell colonies, often referred to as multicellular spheroids (MCS). These two scaffold-free 3D constructs can potentially generate their own ECM components.

Holtfreter and Moscona demonstrated the first formation of MCS using self-assembled cells suspension without external forced interaction with a biomaterial [8]. With this technology, MCS became an important 3D model for tissue engineering and drug testing. A multicellular model is attractive because of its simplicity and ability to mimic the native tissue with a closely packed heterogeneous cell population. Compared to a 2D cell culture, MCS poses improved growth kinetics, better biochemical signalling and enhanced physiochemical gradient. Typical MCS generation methods are cell culture on non-adherent surfaces, spinner flasks, rotating reactor and microwell arrays. Despite the advantages mentioned above, conventional methods for growing MCSs have limited performance in terms of standardized reproducibility and size uniformity. Spheroids produced from conventional methods are usually transferred to another platform for functional characterization and drug testing. This process is often laborious and affects the quality of the spheroids. A microfluidic device can provide a solution for this bottleneck, allowing for high-throughput generation and handling of spheroids. The high-throughput platform is an extremely attractive approach for the clinical applications such as preclinical and therapeutic drug testing.

Since the 1990s, the cutting-edge technology of microfluidics has been adopted for cell culture and producing reliable 3D tissue models that are highly complex, reproducible and tuneable [9]. The small liquid volume, as a key advantage of microfluidics, has been utilized for generating MCSs and the associated cell-based assays. A microfluidic platform is robust and provides several vital features for maintaining in-vivo physiology, such as: (i) integrated components for supplying nutrient and removing waste [10]; (ii) concentration gradient generators suitable for drug delivery and efficacy investigation [11]; (iii) integration of multiple cell-handling tasks such as cell positioning [12], trapping [13] and mixing [14]; (iv) low-cost assays for bioanalysis and high-throughput drug screening [15]; and (v) automated processing to replace tedious manual and robotic handling [16]. Droplet-based microfluidics is an emerging branch of microfluidics that enables the production of highly uniform droplets. This technology allows mixing and encapsulating of cells in a single droplet, which is protected by an immiscible liquid phase. With a high surface area to volume ratio, the microdroplets serve as a unique microbioreactors for a high-yield formation of spheroids [17]. Biomaterials such as polymers and colloid particles were used as the supporting substrate [18]. In addition, droplets containing MCS can be precisely positioned in an array for applications such as cytotoxic testing [19]. Such a droplet array is feasible to control administration of a drug and suitable for high-throughput image-based drug screening.

The present review highlights the recent advances in the development of microfluidics-based 3D spheroid culture. The review focuses on: (i) the formation of MCS in a microfluidic system; (ii) optimization of a microfluidic system for 3D culture; and (iii) integrated systems for high-throughput drug screening. Furthermore, the review also compares and discusses the advantages and limitations of various microfluidic techniques and proposes the future research opportunities, especially to address the current challenges in the field of health care.

2. Multicellular Spheroids

2.1. Formation of Multicellular Spheroids

Tissue compaction and cohesion are essential for the spontaneous formation of cell constructs. An engineered spherical tissue should possess viable cells, organized matrix and biomechanical properties to achieve the ultimate biomimicry. Scaffold-free approaches produce tissues by mimicking natural processes that occur during embryogenesis, morphogenesis and organogenesis. If cells are seeded on a planar surface without exogenous material, they inhibit surface adhesion and form clumps in suspension. The two distinct assembly categories are the organization of extracellular matrix and the self-assembly of cells aggregation. The self-assembly process is spontaneous and more pronounced to intrinsic sorting. The process initiates the self-arrangement and cell–cell interactions to organize aggregates in the form of the spheroids or other shapes. Initially, cells aggregate and undergo self-sorting in response to the signal they generate. The cells subsequently form loose clusters. In addition, the self-sorting mechanism allows cells to selectively form a discrete population and segregate from other populations. A number of intercellular adhesion models have been formulated to describe the formation and compaction of MCS. These models mainly consider the three-step process shown in Figure 1: (i) interaction between ECM and integrin to promote cell attachment (Figure 1A); (ii) up-regulation of cadherin upon cell aggregation (Figure 1B); and (iii) homophilic interaction of type-1 transmembrane proteins (E-cadherins) to initiate strong cell adhesion (Figure 1C).

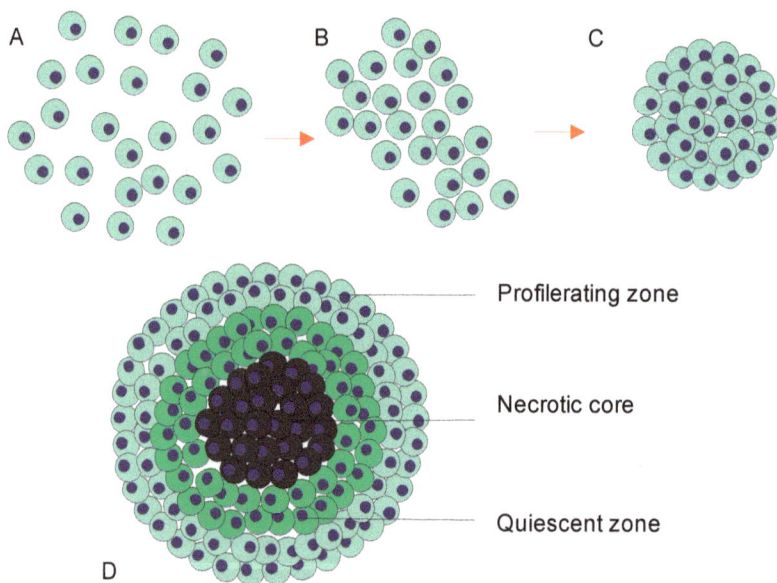

Figure 1. Structure and formation of a multicellular spheroid: (**A–C**) formation; and (**D**) structure.

The interaction between ECM and integrin plays an important role as physical linkers to mediate the cells merging process. A long-chain polymerized fibronectin matrix is the major ECM component that links the cells [20]. Mono-dispersed cells make cell–cell contact and enhance ECM generation with multiple tripeptide arginyl-glycyl-aspartic (RGD) motifs. The RGD motifs then bind with integrins to form integrin/ECM fiber-mediated composition on the cell membrane surface [21]. The integrin–ECM interaction is the base for cell binding that eventually promotes a stronger cell–cell adhesion which is essential for the acceleration of cell aggregation (Figure 1A). Subsequently, clusters of the cell aggregates fuse and assemble into a loosely assembled spherical cell structure. This loose aggregate is highly permeable to nutrient and soluble factors. The next process is known as delayed process in which cells establish the cohesion activity. The integration of the integrin–ECM facilitates this process to increase the cell aggregation. Cells in the aggregates are healthy and possess higher survivability [22]. The regulation of cell–cell recognition and interaction at the delayed phase is initiated by E-cadherin (Figure 1B) [23]. The classical role of cadherins is to increase the adhesiveness of the cell–cell contact and provide a structural support.

E-cadherin expression is established to initiate the cell compaction and it is auto regulated until a specific threshold [24]. Furthermore, cadherin molecules bind with each other by homophilic interaction to generate a strong cell cohesion which is connected by adherent junctions [25,26]. Finally, the loose aggregates compact and condense into a spherical tissue, forming a multicellular spheroid. E-cadherin mediated adhesion is an essential factor for the plasticity of cells which allows the cells shape to change upon contraction (Figure 1C). In brief, co-localisation between the E-cadherin and cytoskeleton facilitates the actin filament rearrangement to form a bundles and networks [27].

2.2. Properties of Multicellular Spheroid

The cells within a MCS are heterogeneously exposed to the nutrient, growth factors and oxygen supply due to MCS structure. Thus, it is meaningful to engineer MCSs with a pre-defined size to be able to predict and regulate the self-sufficiency of MCSs for survival and homeostasis. The metabolic activities within the MCSs are heavily depended on their ability to maintain a sufficient degree of mass transport. [28]. A relatively thick MCSs with diameters ranging between 150 and 200 µm is typically composed of diffusion-restricted tissue. Oxygen diffusivity in a MCS is one of the critical factor which is commonly measured using O_2-sensitive microelectrodes [29]. It is important to note that the zone next to the core of a MCS is most likely to have insufficient oxygen supply and therefore it is more susceptible to the decrease in metabolic rate. Moreover, diffusion of nutrient is another hurdle. Alvarez-Perez et al. reported that proton magnetic resonance with pH-sensitive indicator shows a poor diffusion of nutrients within the MCSs [30]. Furthermore, the cell barrier impedes the elimination of waste, and therefore dumps them at the core of the spheroid. The cells at this region lose their biological activity leads to the occurrence cell death and formation of necrotic core [31]. Figure 1D shows the three basic layers of the MCSs: (i) the outer layer containing proliferating cells with active metabolic rate; (ii) the middle layer consisting of cells at quiescent state in which cells may potentially attain higher metabolic rate and proliferate upon exposure to nutrient; and (iii) the inner region containing cells underwent necrosis. Cells at this region suffer from insufficient nutrient and lose their biological activity to excrete waste.

Considering their structure, MCSs seem to be replicating the architecture of a native tumour tissue. In nature, tumour microenvironment in vivo is biologically heterogeneous, comprising of progressive growth and induce metastasis by extensive neovascularization called "tumour angiogenesis" [32]. The angiogenesis in the tumour predominates the cells survival and proliferation by promoting the biochemical mass transport. Unexpectedly, an increase in angiogenesis decreases the drugs response to the therapy. Realizing this, it will be beneficial to generate co-cultured MCS tumour model by using cancer cells and endothelial cells which will ensures heterotypic cell–cell interaction to form a MSC with tumour vascularity [33]. The vascularized tumour spheroid closely resembles the cellular heterogeneity of solid in vivo tumours.

3. Conventional Methods for Spheroid Generation

Recently, a wide range of basic and complex methods has been developed to generate MCSs. The most important prerequisite is the ability to control the size and uniformity of MSC formation to maintaining optimal biological functions of the MSCs. The grand challenge for MSC generation is maximizing the cell density in a small volume and generating a spheroids with the size smaller than 150 µm [34]. The small size facilitates the homogenous delivery of oxygen and nutrients from the exterior to the core. Furthermore, the culture method should enable the efficient drugs delivery and ideally mimic the in-vivo efficacy [35]. On the other hand, it is crucial to establish methods with standardization, automation, quality control and validation. Such initiatives will potentially accelerate the large-scale production of well-defined MSC for drug screening purposes. In summary, adaptation to high throughput formats will be cost effective. Some of the conventional methods for the production of MSC are described as follow.

3.1. Pellet Culture

Pellet culture is a simple and rapid approach, which was first introduced by Kato et al. [36]. The method employs centrifugal forces to maximize the cell–cell contact and subsequent adhesion at the bottom of a test tube (Figure 2A) with the typical centrifugation acceleration and time as 500 g and 5 min respectively. However, the major drawback of the method is that the shear stress from centrifugation may damage cells and thus provide unreliable results. Another drawback is that this method creates relatively larger spheroids with the diameter of more than 500 μm. Larger MSCs with the diameter range of 500 μm are not suitable for general bioassay development, because oxygen demand in such a large MSC causes hypoxia in the core and thus may not reflect the treatment effectiveness [37]. However, this model is suitable for studying bone regeneration [38]. Typically, low oxygen environment stimulates the differentiation of chondrocytes or chondrogenesis of mesenchymal stem cells [39,40]. Furthermore, this method is not scaled for mass production for high throughput screening and image analysis.

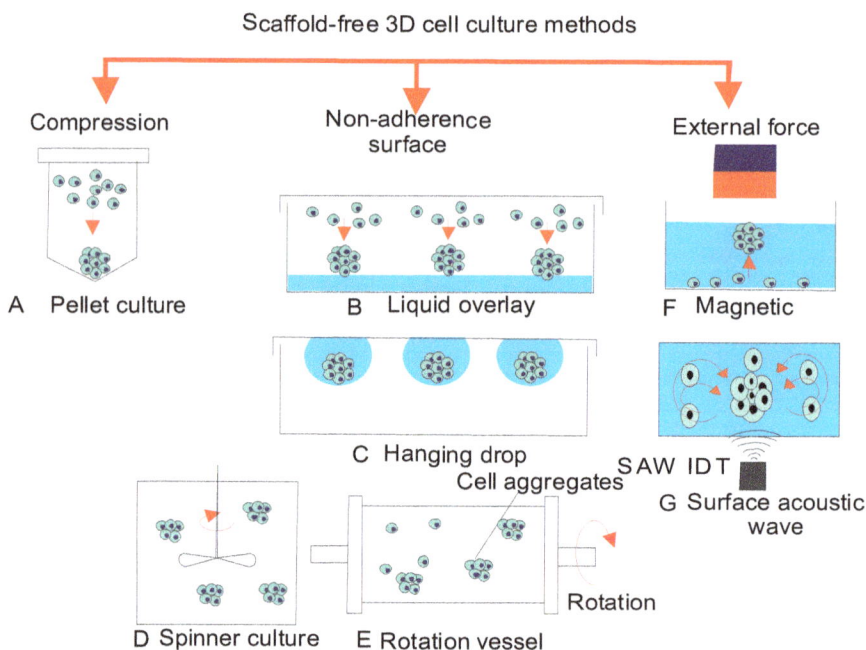

Figure 2. Conventional methods for spheroid generation: (**A**) pellet culture; (**B**) liquid overlay; (**C**) hanging drop; (**D**) spinner culture; (**E**) rotating vessel; (**F**) magnetic force; and (**G**) surface acoustic wave.

3.2. Liquid Overlay

The liquid overlay method allows the cell–cell aggregation instead of the cell adherence by coating the cell culture plate with a non-adherence layer (Figure 2B). This technique was established by Ivascu and Kubbies [41]. A low-adhesive surface was created by coating poly (2-hydroxethyl methacrylate) (pHEMA) on commercially available plates designed with V-shaped or U-shaped bottoms [42]. After seeding cells, a low magnitude of mechanical vibration was applied to promote cell aggregation leading to the MSCs formation. This method is straightforward and easy but suffers from the issues such as reproducibility with sufficiently high yield [43] and non-uniform shape of the spheroids [44]. The usage of different types of culture plates is the main cause for the insufficient reproducibility in size and shape. However, the utility of commercially available Cellstar®Cell-Repellent Surface well plate had shown a high performance [45].

3.3. Hanging Drop

The conventional hanging drop culture was first described by Keller for initiating the development of embryonic bodies [46]. This method is based on the sedimentation of the cells due to gravitational force which promotes the cell–cell interaction. These interactions dominate over cell-substrate interaction, leading to the formation of spheroids (Figure 2C). A small droplet containing cell suspension with volume ranging from 20 to 30 µL is seeded onto the lid of polystyrene microwell plate. After turning the plate upside down, the droplets hang and gravity allows the cells to settle at the bottom for self-assembly. This method simplifies the laborious liquid handling processes and has the potential for high throughput. For instance, Tung et al. proposed a 384-well array to generate hanging drops for high-throughput screening [47]. The droplet maintains the shape and rigidly attached to the lid due to the surface tension. The liquid–air interface of the droplet allows gas exchange but is subjected to extensive evaporation [48]. To prevent evaporation, a modified liquid bath reservoir was used to maintain the humidity.

The hanging drop method is applicable for a volume of less than 50 µL because a large volume reduces the role of surface tension against gravity, and the droplet may fall down [49]. Furthermore, owing to the small volumes and the insufficient nutrient supply, a hanging drop cannot sustain a long-term culture. Changing medium by multiple pipetting steps is susceptible to mechanical perturbation which leads to the spreading or collapse of the droplet. Thus, it is not an easy task to add drug or soluble factors to the droplets. However, Frey et al. developed a hanging drop method incorporating a continuous liquid flow network which enables nutrient supply [50]. Hsiao et al. modified the microwell plate with a lid containing ring structures to improve the structural integrity of the hanging drop [51]. The plate is designed with an extra ring for holding more liquid and is more robust against evaporation. Nonetheless, overall conventional hanging drop techniques do not allow real-time imaging to track the formation process of the spheroids. To overcome this limitation, the primary plate of hanging drops are transferred to a secondary plate using transfer and imaging (TRIM) plates [52] or commercially available systems such as InSphero GravityTRAP system.

3.4. Spinning Flasks

Spinning flasks culture is a cell agitation approach based on stirred suspensions. An impeller mixer in the reactor tank prevents cells sedimentation and also promotes cell–cell interaction in the culture medium, Figure 2D. The spinning mechanism allows for sufficient supply of nutrients and soluble factors to cells while facilitating the excretion of wastes [53]. This method is suitable for long term culture. However, the major limitation is the production of non-uniform spheroids. Furthermore, method needs the specialized equipment and consumes a large amount of medium (100–300 mL). Finally, for drug screening purposes, this method requires manual selection of spheroids to obtain a population with similar size [54].

3.5. Rotating Vessels

Rotating cell culture bioreactors or rotating wall vessel (RWV) was developed by NASA in 1992 to study cell growth under simulated microgravity condition [55]. Instead of stirring, rotary bioreactor rotates itself to maintain cells in a continuous suspension (Figure 2E). The culture chamber rotates along the horizontal axis to prevent cells to adherence to the chamber wall. The speed of the rotation can be adjusted to exert low shear force, and at the same time to promote the optimal cell–cell adherence for a larger 3D structure. Once 3D aggregates are formed, continuous rotation at a desired speed prevents coalescence of the spheroids. This method enables the development of spheroids with approximately uniform sizes. Rotating vessel method ideally suits for long-term culture of 3D spheroids, because easy replacement of medium allows for efficient supply of nutrient and removal of waste [56]. The main disadvantage of this method is the requirement of specialized equipment.

3.6. External Forces

The application of external forces has long been employed to induce cell aggregation and compaction to form 3D structure. Widely used actuation concepts are dielectrophoresis [57], magnetism [58] and acoustic waves [59]. However, these methods are fairly complex, offer little access for visualization of spheroid formation and require specialized equipment. Furthermore, cells are forced to aggregate into a non-uniform geometry, which leads to a heterogeneous morphology and lower yield. These methods can induce mechanical stress onto the cells and may lead to cell damage. Therefore, the formation of spheroids with external forces is not well suited for drug screening. Nevertheless, this approach can meet the demand for 3D culture particularly for cells with lower adherence properties. Recently, magnetic levitation and surface-acoustic-wave have gained increasing attention for the formation and manipulation of spheroids.

3.6.1. Magnetic Levitation

Magnetic levitation was successfully used to construct 3D in-vitro models. This method was used to culture various cell types such as adipocytes, vascular smooth muscle cells and breast tumour cells. Engineering the cell composition and density allows for the formation of the heterogeneous spheroids (Figure 2F). Magnetic levitation was reported to enable tumour and fibroblast cells to interact and form larger spheroids in a shorter time [60]. Moreover, larger spheroids with necrotic cores and region of hypoxia are similar to the in-vivo tumour niche which makes them well suited for the cancer studies. In the study reported by Haisler et al., magnetic nanoparticles were inserted into the layer of 2D confluent cells and then levitated them with an external magnetic field. [61] The magnetic force promotes cell aggregates at the air-liquid interface. The aggregated cell clusters naturally trigger cell–cell interactions. The magnetic nanoparticles are biocompatible and do not induce inflammation or affect the cellular physiology [62]. With this method, cells exhibit improved growth condition, followed by the formation of ECM and compaction. The cohesive multicellular assembly lasted 72 h before forming a spherical shape with a maximum diameter of 1 mm [63]. However, the potential effect of magnetic particles on cellular physiology and metabolism is still not clear and well understood.

3.6.2. Acoustic Wave

Ultrasonic manipulation can be used to concentrate and trap cells in suspension. Bazou et al. observed the changes in the cytoskeleton and adhesion of molecules after cells were exposed to an ultrasound standing wave trap (USWT) (Figure 2G) [64]. Further, a similar approach was employed by Liu et al. to obtain the formation of 3D aggregates of hepatocarcinoma cells. This was followed by the rapid changes in intracellular F-actin within just 30 min [65]. This method is a contactless and can induce larger cell aggregates. The acoustic method provides excellent biocompatibility and contamination-free conditions. This method can be further developed into a tuneable tool to generate MCS. By altering the frequency of the standing wave, spheroids are dynamically assembled and fused into a larger organoids or a desired tissue without using a mould or a template [66]. Controlling the standing surface acoustic wave enables the formation of an acoustic tweezer to precisely pick and assemble cells into an organized 3D structure [67]. More recently, 3D acoustic tweezers were utilized to control and adjust the geometry of spheroid production in only 30 min, achieving a high throughput [68].

4. Microfluidic Methods

Microfluidic technology has rapidly evolved in biomedical research as a powerful tool for various applications such as cell-based assay, tissue engineering, molecular diagnostics and drug screening. The basic tasks of microfluidic technology are processing and manipulating small amounts of liquid (10^{-9} to 10^{-18} litters); in a size scale that matches the size of cells and microtissue. Microfluidics is categorized as continuous-flow and digital microfluidics (Figure 3). Continuous-flow microfluidics can be further categorized as single-phase and multi-phase microfluidics. Multi-phase microfluidics and digital microfluidics can accurately and efficiently produce micro droplets within milliseconds enabling a high throughput [69] for cell-based analysis [70]. The implementation of 3D cultures in microscale potentially allows further reduction of the volume of nutrient, reagents, soluble factors and drugs. 3D cell culture with microfluidics provides a higher controllability and provides a cost-effective mean for biomanufacturing. Furthermore, droplet-based microfluidics offers the advantages for cell compartmentalization with a high surface-to-volume ratio. This physical characteristic is desirable in a wide range of molecular and cellular analysis [71]. As the viability of cells during the generation of MCS is critical, a microfluidic platform typically applies a low shear stress and thus minimizes cell damage [72]. Furthermore, an integrated microfluidic device provides a high degree of programmability and configurability. For instance, advance cell-based assays are scalable as an array. This feature allows microfluidics not only to engineer microscale 3D tissues but also more complex tissue system such as artificial organs on a chip. In the following sections, we discuss the state of the art of microfluidic technology for MCS generation and analysis including: (i) platform configuration and applicability; (ii) integrated microfluidics; and (iii) technical limitations and improvement.

4.1. Continous-Flow Microfluidics

A continuous flow in microchannels is either delivered by a flow-rate-driven or pressure-driven pumping system. Fluid flow in this small scale is inlaminar regime as surface effects such as friction dominate over volume effects such as inertia. Continuous-flow microfluidics either handles a single-phase or multi-phase segmented flow. Multi-phase flow microfluidics is also often called a droplet-based microfluidics, which generates and manipulates monodispersed microdroplets.

In single-phase continuous-flow microfluidics, the microchannels are coated or filled with hydrogel for trapping and providing a scaffold for cell growth. The technology provides a precise concentration gradient of soluble molecules, nutrients and drugs [73]. Thus, an artificial microenvironment can be created to facilitate spheroid growth (Figure 3A) [74]. In multi-phase continuous-flow microfluidics, liquid droplets are formed and manipulated in a continuous manner. The two established methods for droplet formation are flow-focusing and T-junction configurations (Figure 3B–D). The flow-focusing configuration forms microdroplets by squeezing the liquid stream with two immiscible sheath streams to generate highly monodisperse droplets [75]. The T-junction configuration uses a single sheath flow to break up the dispersed phase into droplets [76]. In both configurations, scaffold materials such as hydrogel can be added into the droplets to support cell growth.

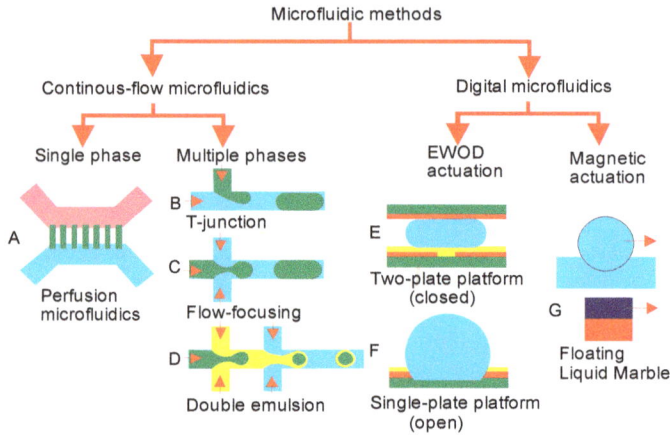

Figure 3. Microfluidics based methods for the generation of spheroids. Single-phase microfluidics: (**A**) perfusion microfluidics; Multi-phase microfluidics; (**B**) T-junction; (**C**) flow-focusing; (**D**) double emulsion—electrowetting on dielectric actuation; (**E**) two-plate platform; (**F**) single-plate platform—magnetic actuation; and (**G**) floating liquid marble.

4.1.1. Single-Phase Microfluidics

Single-phase microfluidic synergies the 3D cell growth and allow the exchange of media by perfusion system. In principle, fluid is continuously flown within a microchannels similar to in-vivo vascularization effect [77]. Conventional systems for spheroid formation and growth is confined to a reactor and isolated from the surrounding. Consequently, spheroid cultivation is limited to short-term culture and is detrimental to the cell viability. This drawback also exists in droplet-based microfluidics, which insufficiently regulates the environment for spheroid growth. To circumvent these limitations, Agastin et al. attempt to grow multiple tumour spheroids using polydimethylsiloxane (PDMS) microbubbles. A physiological flow was established inside the microbubble by media perfusion [78]. This model mimics an in vivo avascular tumour condition.

Ideally, perfusion method is regarded as a promising platform for anti-cancer drug testing simply because it enhances drug exposure as well as the exchange of nutrients and wastes. In this microfluidic system, media is flown continuously through microchannels and perfuse trapped cells or spheroids. [74]. Thus, the formation and growth of spheroid can be carried out over a long time period (e.g., two weeks) without significant decrease in the cell viability [79]. Moreover, the method allows the possibility of device fabrication with integrated concentration gradient generator for high-throughput applications. [80]. Perfusion flow can be generated simply by gravity and surface tension. In order to increase the throughput, the assay can be scaled up to the 96 wells format. The initial attempt was carried out by Chen et al. developing an integrated micro-capillary network that is connected with supply chambers of culture medium [81]. Following these advances, Sakai et al. designed an improved microwell array with perfused flows to carry out spheroid-based high-throughput drug testing [82]. Another interesting approach was to combine multi-phase and single-phase microfluidics. Spheroids are initially generated in an emulsified droplet, then by displacement of surfactants droplets lead to coalescence and spheroids are exposed to perfusion [83]. Data acquisition and analysis from chemotherapeutic studies showed perfusion-based format showed higher chemoresistivity as compared to the static fluidic environment [84]. Interestingly, this system also allows miniaturizing the combination of both physiologic and pathologic networks such as angiogenesis and thrombosis in a 3D capillary network [85]. For instance, the platform could serve as a model for pathophysiological conditions such as tumour angiogenesis and tissue ischemic model.

4.1.2. Multi-Phase Microfluidics

In the past five years, several reviews presented a comprehensive overview of methods to form, sort, and merge and manipulate microdroplets for experiments in chemistry and biology [86–88]. Recently, droplet-based microfluidics becomes an attractive approach as a microbioreactor to grow and characterize living cells [89] and protocells [90]. Water-in-oil droplets serve as vessels for cell culture. A continuous aqueous flow breaks into droplets and encapsulated by an immiscible phase such as mineral oil with biocompatible surfactants [91]. The immiscible oil phase (W/O), droplets are not optimal for the generation of MCS because they impedes the supply of nutrient and gas exchange [92]. Thus, droplet-based systems only allow for a short-term cell culture. To circumvent these problems, a water-in-oil-in-water (W/O/W) double emulsions (DE) format was used to encapsulate cells. These droplets function as selective barrier to regulate the transport of soluble factors and nutrient. Moreover, the outer aqueous phase ensures the adequate permeability of the oxygen. However, conventional DE generates highly polydisperse droplets which are easily breakable. [93]. The precise control of the surface wettability in the device is critical for the stability of the droplets [94]. The droplet also improves heat and mass transfer and increases the reaction rate. These advantages accelerate the diagnostic results as it allows the molecular or enzymatic reaction to occur in a shorter period of time. Further in the next section, the use of single-phase and multi-phase microfluidics for 3D cell culture is discussed in detail.

The two main challenges of droplet-based microfluidics are: (i) the formation of biocompatible, monodisperse (<1%–3% dispersity) and stable droplets; and (ii) the real-time observation of cell activity. The size and stability of droplet is critical to sustain long term culture of MCSs as well as drug testing. The hydrodynamic properties such as type of flow, laminar or turbulent, are crucial for scaling of the size of the droplets during emulsification process [95]. The emulsion quality also depends on the viscous shear stress and interfacial tension. The breakup of droplets occurs when viscous shear stress dominates and overcomes the interfacial tension. Further, when generated in bulk, shear force and inertia cause the droplet to coalesce or break [96]. To overcome these limitations, a biocompatible amphiphilic molecule called "surfactant" is absorbed at the interface. Surfactants reduce the interfacial tension between the dispersed and continuous phases, which is critical to maintain the droplet stability and prevent coalescence [97].

The outer oil layer of a droplet serves as a selectively permeable barrier and allows the transport of small molecule across this barrier, which is essential to provide a discrete microenvironment for cell culture. To further enhance cell growth the droplets can be formulated by encapsulating cells with biological additives. Tumarkin et al. demonstrated the possibility of using microgel-based biomaterials, which potentially promoted the cell functionality in terms of enhanced proliferation and adhesiveness for the formation of MCS [98]. For instance, a continuous-flow microfluidic system can be configured to co-encapsulate cells into hydrogel. The gelation process forms the cell-laden microcapsules. Hydrogels are crosslinked with chemical residues by chemical stimuli and also with physical process such as radical reactions, temperature and photon energy [99]. Hydrogel gel particles provide structural support for spheroid growth and function [100,101]. The hydrogel capsules can be scaled up as spheroid carriers to serve as immunoisolation barrier for the cell transplantation [102]. A variety of chemically modified hydrogels have been used to make microcapsules such as alginate-PLL (poly-Llysine)-alginate (APA), thermally responsive hydrogels (agarose, NIPAM based hydrogel and gelatin) and photosensitive hydrogels such as polyethylene glycol (PEG). The precise mechanism of the various types of the encapsulation including cells types, functional outcomes and limitations are summarized is Table 1.

Table 1. Encapsulation types for 3D cell culture in droplets.

Polymer	Gelation Method	Advantage	Disadvantage
Agarose	Temperature Shift	Improved nutrient diffusion [103] Biocompatible [104]	The gelling temperate must be conducive for optimal cell viability
Gelatin	UV irradiation	Formation cell–matrix interactions with hydrogel [105]	Combination with hydrogel liquefies [106]
Poly(ethylene glycol) (PEG)	UV irradiation	Biodegradable [107]	Poor drug release [107]
lactic-co-glycolic acid (PLGA)	UV irradiation	Release hydrophobic drugs [108]	Poor drug encapsulation [109]
PEG-PLA incorporation	UV irradiation	Ideal for drug delivery [110]	Poor stability [110]
Alginate	Ion reaction	Highly permeable structure and allows long term culture [111]	Rapid gelation process can from non-spherical particles [112]
Pura matrix Hydrogel	Ion reaction	Increase cell attachment, proliferation and differentiation [99]	Decrease the cell viability [113]
Gelatin + Matrigel	Ion reaction	Facilitate cell-assembly [114,115]	The matrigel can induce morphology alteration of the cells [114]

Biomimetic or biodegradable material based microcarriers are useful for cell culture and drug delivery. Shi et al. modified the method of double emulsion solvent evaporation to fabricate a biodegradable Poly(D,L-lactide) porous microspheres [116]. These microspheres act as a microcarrier to deliver cells which could be used for cell-based therapies. In another study, a double emulsion template was used to encapsulate droplet carrying multiple biomimetic scaffold. The scaffolds consist of the porous cores which acted as a microcarrier and provided a confine environment for growing spheroids. [117]. More recently, a reusable device was developed to customize monodisperse droplets and perform multiple droplet encapsulations [118]. Collectively, it can be concluded that, multiple encapsulation technique holds wider application in a 3D cell based assay.

4.2. Digital Microfluidics

A simple platform with easy liquid-handling is critical for lab-on-chip technology. Continuous-flow microfluidics platforms require pumps and tubing for fluid delivery. Thus, handling discrete droplets has advantages over the continuous flow microfluidics. Furthermore, actuation techniques such as magnetic or electric forces are widely used for handling droplets. The digital microfluidics (DMF) is a branch of microfluidics, which integrates the microfluidic devices and electrical forces to manipulate discrete droplets [119,120]. There are two common configurations of DMF for culturing: (i) closed format with droplets sandwiched between two plates (Figure 3E); and (ii) open format with droplets positioned on top of a planar surface (Figure 3F) [121]. The bottom plate usually has an array of actuation electrodes. The top plate made of transparent conductive material allows the optical imaging. For droplet movement with low friction, the surfaces of both plates are coated with hydrophobic material. Droplet manipulation tasks such as dispensing, splitting, merging and coalescence are carried out with electrostatic force by tuning the electric potential of the electrodes [122]. The mobility of the droplet is controlled through a combination of electro wetting (wetting behaviour of liquid) and liquid dielectrophoretic forces (effect of non-uniform electric field on liquid). Thus, DMF offers several advantages over conventional microfluidics such as low cost, portability and low reagent usage, yet provides faster test result. The most significant advantage is the ability to perform multiple biochemical assays simultaneously using a planar array of electrodes [123]. This technology enables the evaluation of several test results in real time.

However, common fluid operation using DMF is restricted to a 2D platform, and therefore it has limitations such as cross contamination, solute adsorption and degradation of soluble factors [124]. Interestingly, the transition from 2D to 3D DMF was demonstrated by submerging droplet in oil between two electrodes. The droplet can be manipulated by moving horizontally and vertically, thus serving as a programmable hanging droplet inside the oil. This platform was successfully used to grow mouse fibroblast [125]. Apart from that, a 3D scaffold based DMF platform was employed by Fiddes et al. to culture NIH-3T3 cells in hydrogel discs [126]. Subsequently, Au et al. improved the scaffold based-DMF to grow HepG2 and NIH-3T3 co-cultured spheroids (organoids) using collagen hydrogels. This method was applied for to hepatotoxicity screening [127]. Recently Aijian et al. demonstrated the adoption of the digital microfluidic in a hanging drop based platform for the generation of spheroids [128]. This method permits automation of liquid handling by dispensing liquid through connected wells to form hanging droplets. This platform is automatable and flexible for liquid handling which enables the formation of hanging drop. These sequential and reconfigurable operations increase throughput for spheroid based assays.

Our recent approach for multiple spheroids of olfactory ensheathing cells was carried out using floating liquid marble (LM) [129]. A drop of liquid was coated with hydrophobic powder to form an elastic hydrophobic shell with fine pores allowing the exchange of gas. LM also has a low evaporation rate and regulates humidity in floating condition. Moreover, the floating mechanism eases the cell interaction inside the LM due to the internal fluid flow. Additionally, the liquid marble suits as a microbioreactor to generate and to differentiate embroid bodies [130]. Given its advantage of miniaturization, the LM platform allows for the adoption for high-throughput drug screening [131]. Recently, Ooi et al. reported that LM containing ethanol exert self-propelling caused by Marangoni solutocapillary effect [132]. It is also possible to control the locomotion of floating LM by using magnetic actuation (Figure 3G). In the study by Khaw et al., magnetic particles were added to the LM and a moving permanent magnet was used to drag the LM [133]. The actuation of LM is crucial and brilliant for engineering controllable and tuneable functions. LM composed with magnetic nanoparticles can generate centrifugal force to function as microcentrifuge [134]. Additionally, another study addresses the usage of magnetism to split LM with lycopodium—iron oxide [135]. In summary, it is evident that LM can be utilized to support a digital microfluidic platform for 3D cell-based application.

The combination of microfluidics and optics results in a unique technology called optofluidics. The initial optical applications in microfluidic domain were optical tweezers [136] and optical vortex [137]. Photoconductivity can be applied to conventional (DMF) to actuate droplets [138,139] using opto-electro-wetting (OEW) [140]. The technology has shown potential for achieving higher control accuracy by combining photosensitive surfactants and the laser. More recently, 3D droplet manipulations was demonstrated by fabricating a single-sided continuous opto-electro-wetting (SCOEW) platform, which is supposed to have advantage over conventional EWOD and OEW [141]. Further, digitalizing and integrating of the microfluidics with optofluidics offers a powerful imaging solution for biomedical imaging. For instance, the digital holographic microscope is lens free and facilitates real-time imaging three-dimensional tomography imaging of transparent PDMS opto-microfluidic channel [142]. Efforts to utilize this technology with microfluidic include 3D sensing of microorganism [143] and automated cell viability detector [144]. However, application for 3D cell culture has not been yet explored.

5. Application of Spheroids in Microfluidic

5.1. Organ Printing

Organ printing is promising to transform tissue engineering into customized organ biofabrication. Spheroids make an excellent candidate for organ bioassembly. A spheroid has an ideal geometry and may serve as "bioink" for bioprinting. Printing the spheroids layer-by-layer for tissue constructions is a common approach in biopriniting. With the advance computer-aided robotic bioprinting technology, the predefined structure can be precisely printed to achieve desired organ/tissue assembly. The positioning and the placement of the dispersed spheroids are critical factors to achieve a controllable fusion in a 3D tissue. Most recently, Moldovan et al. reported a latest invention called the "Kenzan" method, which utilises microneedles for the spheroids assembly [145]. This technique holds the precision up to micron-level and is able to link the spheroid closely. Further, the array of tightly aligned spheroid undergoes a fusion process to form a complex tissue and syntheses their own ECMs. Tissue construction using spheroid requires a large quantity of uniformly sized spheroids which is required to achieve bio printing with satisfactory resolution. Importantly, a scalable spheroid fabrication method is crucial for producing large quantity of homogenous spheroids. Ultimately, a precise 3D tissue print is achievable by the use of microfluidic based spheroid bio fabricator. Further, application of droplet based digital microfluidic may offer a scalable production of the spheroids at high yield [146]. Figure 4 describes the integration of microfluidic technology for spheroid based 3D printing.

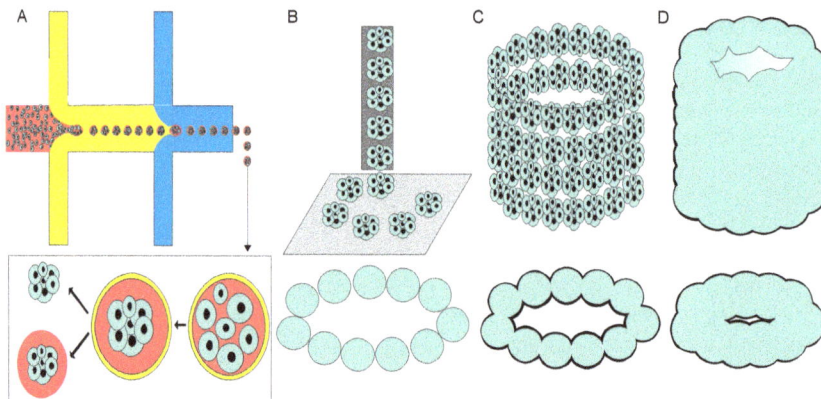

Figure 4. Schematic presentations of principles of 3D tissue spheroids printing: (**A**) microfludic based fabrication of spheroid; (**B**) nozzle is used to dispense spheroids; (**C**) continuous dispensing form layer-by-layer tissue spheroid; and (**D**) layer-by-layer tissue spheroid fusion and bio-assembly of tubular tissue construct.

Many other technologies can be considered to improve the use of spheroids for bioprinting as well as tissue fabrication. For instance, a bioprintable scaffold such as electrospun matrix can accurately pattern printed spheroids into desired tissue construct [147]. Furthermore, magnetic 3D printing has the potential to achieve a precise and rapid construction of 3D tissue. For example, a recent study reported a spheroid patterning technique using superparamagnetic iron oxide nanoparticles (SPIONs) for bioprinting [148]. Another approach is using magnetic levitation to construct tumour spheroids which closely mimic native microenvironment [149]. The mechanism of spheroid fusion involves 3D cell–cell interaction and critical for the formation of larger tissue. To date, only a few studies revealed the mechanism of spheroid fusion. Quantification of fusion kinetics, accounting the time lapse for the coalesce of two spheroids is necessary [150]. Recently, Munaz et al. demonstrated a microfluidic platform which can be used to study spheroid fusion as well as drug screening to promote fusion [151]. However, more sophisticated platforms are needed to quantify the cellular parameters such as mechanical strength of fused spheroids.

5.2. Organ-On-Chip

To date, the microfluidics-based lab-on-a-chip technologies are facing challenges to reduce the costs and increase the efficiency for drug screening and development. However, with further development it may potentially serve as a tool for preclinical models for human efficacy and safety. Prominently, this technology contributes towards an alternative step to reduce the extensive use of animal testing. Further, the technology allows for mimicking the complexity of animal-based testing models. The combination of fluid physics with 3D cell compartmentalization has gained popularity as organ-on-chip devices. Interestingly, the organ-on-chip concept simplifies clinical bioanalysis by integrating realistic organ models in a single device [152,153].

Initially, the organ-on-a-chip concept was established by combining the cultures of the liver spheroid and neurospheres in a separate chamber and connecting them by a microfluidic circuit [154].

Subsequently, a scaled-up organ-on-chip device was fabricated by combining spheroids grown in arrays and perfused with culture media, which resulted in spheroids fusion and tissue formation [152]. These tissues may potentially serve as a model to simplify physiological function of an organ. The advanced organs-on-a-chip models are integrated with microsensors, which can detect cells and environmental cues. For instance, to measure the transmembrane electrical resistance across cell barrier and detecting the cell migration [155]. In the near future, this technology can be used for biochemical analysis and biophysical analysis (i.e., tissue mechanics, invasions and fusion). Furthermore, it may also facilitate the testing on molecular diagnostic (i.e., protein, nucleic acid detection).

Many biological processes involve the interaction between multiple organs. Thus, a futuristic organ-on-a-chip model will incorporate multiple organ interactions in physiologically relevant orders. This technology is likely to have a higher accuracy for testing drug metabolism and toxicity for translating basic bench research to clinical practice. This device can be fabricated by using spheroids to further mimic the in-vivo conations of cells and enhance the results [156]. This approach mimics the multiple organ systems in a single device which is termed as a "body-on-a-chip". Furthermore, the model is designed with a fluid stream, which acts as a network of a surrogate blood vessel to interconnect all tissue compartments biochemically. Thus, the model allows for testing drug effect based on multiple physiological interactions. In the context of drug testing applications, this platform is promising and may increase the screening throughput and facilitate the development of new drug candidates. Further, it allows bioanalysis to identify the pharmacokinetic and pharmacodynamic consequences to predict the safety of the drugs. Since microfluidic cartridges are translucent, they are also useful for time-lapse imaging to identify drug induced pathophysiological changes. Figure 5 shows a schematic presentation the possible use of spheroids to fabricate multi-organ-on-a-chip.

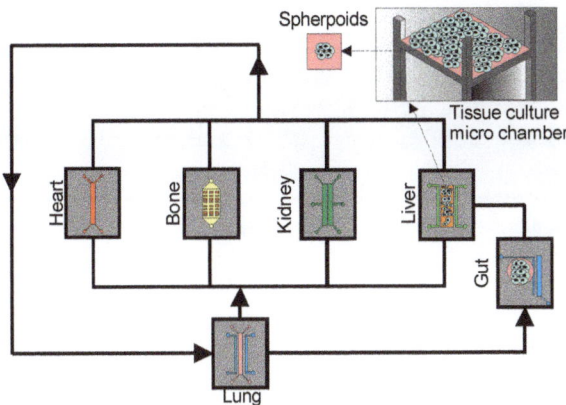

Figure 5. Schematic example of spheroid integration with microfluidic based multiple organ-on-a-chip models.

Organoids on Chip

Both spheroids based "organ-on-a-chip" and even a "body-on-a-chip" is still in infancy to accurately reproduce and mimic the in vivo niche. Although spheroids consist of organized tissues, the quality of tissues function is limited. Thus, the developments towards the use of organoid cultures have gained attention in the recent years. Organoid can be grown onto a microfluidic platform to model organ features such as development, homeostasis and diseases. These cultures form a higher order tissue organization such as hollow spherical tissue. The organoids are generated using stems cells or embryonic stem. The organoids exclusively represent development system of the embryonic tissue. The stem cells potentially aggregate and assemble into a spatially patterned structure that supports organogenesis. Moreover, organoids also can be developed using induced pluripotent stem

(IPS) derived from a patient who represent a personalized cell system that functions as a disease model for an individual.

The organoids are "near physiological" models for drug metabolism and toxicity testing. The development of organoids of the gastrointestinal system also represents an important resource for drug development programs. Initially, intestinal organoids were developed by using stem cells expressing leucine-rich repeat-containing G-protein coupled receptor 5 (Lgr5) [157]. Furthermore, these Lgr5+ stem cells can be differentiated and develop organoids of gastrointestinal system (GI) including hepatic system. The organoids of the hepatic system enable the acquisition of biotransformation drugs and toxins to predict drug safety. Recent advances in a microfluidic system can potentially help to miniaturize the hepatic organoids in a single chip for high-throughput screening. For instance, digital microfluidic system was developed for drug testing using arrays of liver organoids [127]. The platform is known as organoid droplet exchange procedure (GODEP), which allows fluid manipulation, e.g. reagent exchange. A continuous media circulation in microfluidic devices is critical for generating growth factor, signaling and drug gradients. However, it is difficult to maintain organoid position in a continuous-flow microfluidic system. Thus a stationary organoid placement could be beneficial for stabilizing tissue position. Recently, liver organoids were encapsulated and loaded in a perfused C-shaped trap arrays. This approach sustains hepatic tissue position during exposure of various fluid flow rates [158].

Organoids are promising candidates for cellular therapeutics. Microfluidics-assisted encapsulation of organoids provides a prominent role in therapeutic delivery systems. For instance, LSFM4LIFE project is an ongoing EU Horizon 2020 program seeks to achieve cellular therapy for type 1 diabetes by using Human Pancreas Organoids (hPOs). Further, alginate-poly-L-lysine is used to encapsulate cells for secreting antibody and therapeutic protein. These microcapsules are called immunotherapeutic organoids. These organoids can be implanted in vivo to deliver therapeutics. They are alternative to drug based treatment for cancer [159] and immunological disorders [160].

The integration of biomaterial supports the further development of organoid-based platforms. For instance, patterning of hyaluronic acid (HA) as substrate was used to fabricate 3D contractile cardiac organoids [161]. Recently, researchers at Wake Forest School of Medicine have generated functional cardiac tissue by 3D printing spheroids of cardiac organoids. In another study, hepatic organoids were engineered by assembling thousands of spheroids into a predefined pattern using acoustic nodes [66]. Additionally, Shen et al. demonstrated the construction of human airway epithelial by using epithelial organoids in response to cues from an ECM nano patterned substrate [162]. The applications of IPS-derived organoids are summarized in Figure 6.

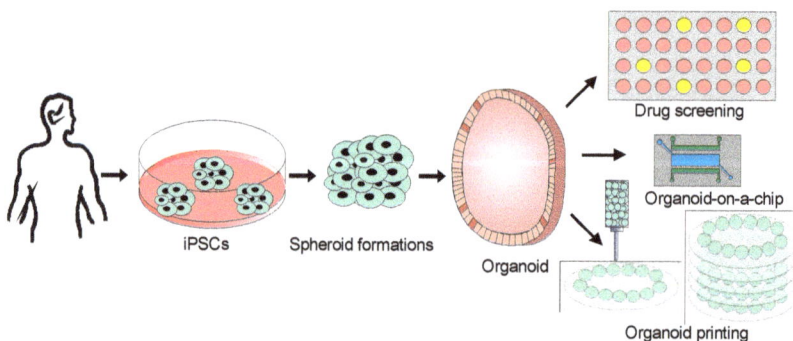

Figure 6. Organoid development and integration with microfluidic technology for drug screening, development organoid-on-a chip model and organoid-based bioprinting.

Micromachines **2017**, *8*, 94

6. Conclusions

In line with the current trend, 3D cell cultures are cost effective and easy-to-use solutions for various applications. The microfluidic technology evidently demonstrated its suitability for the translation of 3D cell spheroid technology into commercial products. In fact, advances in microfluidics-based spheroid culturing techniques made substantial progress in biomedical ventures including drug screening, tissue engineering, disease modelling and cellular therapeutics. However, further optimization is needed to accurately mimic the in-vivo environment. Ideally, physiologically relevant models will help in overcoming the animal testing and animal based assays. Nonetheless, heed should be given to ensure that these microfluidic tools do not interfere or manipulate the ideal cellular behaviour. Ultimately, the usefulness of iPSC, organoids technology and microfluidic system may serve as a breakthrough for featuring next generation human models. Organoids potentially enhance the similarity of 3D tissue organization to that of real organs. Another interesting direction is the utility of 3D-bioprinted organoids for tissue construct. It is also critical to improve efficacy of drug testing by using this technology. In that regard, an advance look on coupling digitalized organ-on-a-chip systems with mass spectrometry analysis will widen the pipeline of future of drug development. Furthermore, it is crucial to integrate new biosensors technology into multi-organ systems. This is particularly important for feedback and molecular diagnostics. Collectively, the future challenges will be scaling up this technology towards the target for personalized medicine. By this, individualized therapeutic methods can be established as more precise medical approaches for intracellular delivery, cell transplantation and personalized tissue engineering.

Acknowledgments: We acknowledge the Australian Research Council for the support through the grant DP170100277.

Author Contributions: N.T.N and M.J.A.S developed the structure of the paper. K.H. and R.K.V. collected and analysed the literature. All authors wrote the paper.

Conflicts of Interest: The authors declare no conflict of interest.

References

1. Mahteme, H.; Lovqvist, A.; Graf, W.; Lundqvist, H.; Carlsson, J.; Sundin, A. Adjuvant 131i-anti-cea-antibody radioimmunotherapy inhibits the development of experimental colonic carcinoma liver metastases. *Anticancer Res.* **1998**, *18*, 843–848. [PubMed]
2. Sanchez-Romero, N.; Schophuizen, C.M.; Gimenez, I.; Masereeuw, R. In vitro systems to study nephropharmacology: 2D versus 3D models. *Eur. J. Pharmacol.* **2016**, *790*, 36–45. [PubMed]
3. Cukierman, E.; Pankov, R.; Stevens, D.R.; Yamada, K.M. Taking cell-matrix adhesions to the third dimension. *Science* **2001**, *294*, 1708–1712. [PubMed]
4. Nath, S.; Devi, G.R. Three-dimensional culture systems in cancer research: Focus on tumor spheroid model. *Pharmacol. Ther.* **2016**, *163*, 94–108. [PubMed]
5. Ashe, H.L.; Briscoe, J. The interpretation of morphogen gradients. *Development* **2006**, *133*, 385–394. [PubMed]
6. Tibbitt, M.W.; Anseth, K.S. Hydrogels as extracellular matrix mimics for 3D cell culture. *Biotechnol. Bioeng.* **2009**, *103*, 655–663. [CrossRef] [PubMed]
7. Liao, J.; Guo, X.; Grande-Allen, K.J.; Kasper, F.K.; Mikos, A.G. Bioactive polymer/extracellular matrix scaffolds fabricated with a flow perfusion bioreactor for cartilage tissue engineering. *Biomaterials* **2010**, *31*, 8911–8920. [CrossRef] [PubMed]
8. Moscona, A.; Moscona, H. The dissociation and aggregation of cells from organ rudiments of the early chick embryo. *J. Anat.* **1952**, *86*, 287. [PubMed]
9. Van Duinen, V.; Trietsch, S.J.; Joore, J.; Vulto, P.; Hankemeier, T. Microfluidic 3D cell culture: From tools to tissue models. *Curr. Opin. Biotechnol.* **2015**, *35*, 118–126. [CrossRef] [PubMed]
10. Fu, C.Y.; Tseng, S.Y.; Yang, S.M.; Hsu, L.; Liu, C.H.; Chang, H.Y. A microfluidic chip with a U-shaped microstructure array for multicellular spheroid formation, culturing and analysis. *Biofabrication* **2014**, *6*, 015009. [CrossRef] [PubMed]

11. Nguyen, N.T.; Shaegh, S.A.; Kashaninejad, N.; Phan, D.T. Design, fabrication and characterization of drug delivery systems based on lab-on-a-chip technology. *Adv. Drug Deliv. Rev.* **2013**, *65*, 1403–1419. [CrossRef] [PubMed]

12. Lin, L.; Chu, Y.S.; Thiery, J.P.; Lim, C.T.; Rodriguez, I. Microfluidic cell trap array for controlled positioning of single cells on adhesive micropatterns. *Lab Chip* **2013**, *13*, 714–721. [PubMed]

13. Zhu, J.; Shang, J.; Olsen, T.; Liu, K.; Brenner, D.; Lin, Q. A mechanically tunable microfluidic cell-trapping device. *Sens. Actuators Phys.* **2014**, *215*, 197–203. [CrossRef]

14. Ainla, A.; Jansson, E.T.; Stepanyants, N.; Orwar, O.; Jesorka, A. A microfluidic pipette for single-cell pharmacology. *Anal. Chem.* **2010**, *82*, 4529–4536. [PubMed]

15. Wen, Y.; Yang, S.T. The future of microfluidic assays in drug development. *Expert. Opin. Drug Discov.* **2008**, *3*, 1237–1253. [PubMed]

16. Ly, J.; Masterman-Smith, M.; Ramakrishnan, R.; Sun, J.; Kokubun, B.; van Dam, R.M. Automated reagent-dispensing system for microfluidic cell biology assays. *J. Lab Autom.* **2013**, *18*, 530–541. [PubMed]

17. Gu, G.Y.; Lee, Y.W.; Chiang, C.C.; Yang, Y.T. A nanoliter microfluidic serial dilution bioreactor. *Biomicrofluidics* **2015**, *9*, 044126. [PubMed]

18. Alessandri, K.; Sarangi, B.R.; Gurchenkov, V.V.; Sinha, B.; Kiessling, T.R.; Fetler, L.; Rico, F.; Scheuring, S.; Lamaze, C.; Simon, A.; et al. Cellular capsules as a tool for multicellular spheroid production and for investigating the mechanics of tumor progression in vitro. *Proc. Natl. Acad. Sci. USA* **2013**, *110*, 14843–14848. [CrossRef] [PubMed]

19. Sabhachandani, P.; Motwani, V.; Cohen, N.; Sarkar, S.; Torchilin, V.; Konry, T. Generation and functional assessment of 3D multicellular spheroids in droplet based microfluidics platform. *Lab Chip* **2016**, *16*, 497–505. [PubMed]

20. Shield, K.; Riley, C.; Quinn, M.A.; Rice, G.E.; Ackland, M.L.; Ahmed, N. $\alpha 2\beta 1$ integrin affects metastatic potential of ovarian carcinoma spheroids by supporting disaggregation and proteolysis. *J. Carcinog.* **2007**, *6*, 11. [PubMed]

21. Loessner, D.; Stok, K.S.; Lutolf, M.P.; Hutmacher, D.W.; Clements, J.A.; Rizzi, S.C. Bioengineered 3D platform to explore cell-ECM interactions and drug resistance of epithelial ovarian cancer cells. *Biomaterials* **2010**, *31*, 8494–8506.

22. Frixen, U.H.; Behrens, J.; Sachs, M.; Eberle, G.; Voss, B.; Warda, A.; Lochner, D.; Birchmeier, W. E-cadherin-mediated cell-cell adhesion prevents invasiveness of human carcinoma cells. *J. Cell Biol.* **1991**, *113*, 173–185. [PubMed]

23. Marsden, M.; DeSimone, D.W. Integrin-ecm interactions regulate cadherin-dependent cell adhesion and are required for convergent extension in xenopus. *Curr. Biol.* **2003**, *13*, 1182–1191. [PubMed]

24. Lin, R.Z.; Chou, L.F.; Chien, C.C.; Chang, H.Y. Dynamic analysis of hepatoma spheroid formation: Roles of e-cadherin and beta1-integrin. *Cell Tissue Res.* **2006**, *324*, 411–422.

25. Takei, R.; Suzuki, D.; Hoshiba, T.; Nagaoka, M.; Seo, S.J.; Cho, C.S.; Akaike, T. Role of e-cadherin molecules in spheroid formation of hepatocytes adhered on galactose-carrying polymer as an artificial asialoglycoprotein model. *Biotechnol. Lett.* **2005**, *27*, 1149–1156. [PubMed]

26. Haga, T.; Uchide, N.; Tugizov, S.; Palefsky, J.M. Role of E-cadherin in the induction of apoptosis of HPV16-positive CaSki cervical cancer cells during multicellular tumor spheroid formation. *Apoptosis* **2008**, *13*, 97–108. [CrossRef] [PubMed]

27. Ayollo, D.V.; Zhitnyak, I.Y.; Vasiliev, J.M.; Gloushankova, N.A. Rearrangements of the actin cytoskeleton and e-cadherin-based adherens junctions caused by neoplasic transformation change cell-cell interactions. *PLoS ONE* **2009**, *4*, e8027.

28. Mueller-Klieser, W. Method for the determination of oxygen consumption rates and diffusion coefficients in multicellular spheroids. *Biophys. J.* **1984**, *46*, 343–348. [PubMed]

29. Mueller-Klieser, W. Microelectrode measurement of oxygen tension distributions in multicellular spheroids cultured in spinner flasks. *Recent Results Cancer Res.* **1984**, *95*, 134–149. [PubMed]

30. Alvarez-Perez, J.; Ballesteros, P.; Cerdan, S. Microscopic images of intraspheroidal pH by 1H magnetic resonance chemical shift imaging of pH sensitive indicators. *MAGMA* **2005**, *18*, 293–301. [CrossRef]

31. Hamilton, G. Multicellular spheroids as an in vitro tumor model. *Cancer Lett.* **1998**, *131*, 29–34. [PubMed]

32. Ghosh, S.; Joshi, M.B.; Ivanov, D.; Feder-Mengus, C.; Spagnoli, G.C.; Martin, I.; Erne, P.; Resink, T.J. Use of multicellular tumor spheroids to dissect endothelial cell–tumor cell interactions: A role for T-cadherin in tumor angiogenesis. *FEBS Lett.* **2007**, *581*, 4523–4528. [CrossRef] [PubMed]
33. Hsiao, A.Y.; Torisawa, Y.S.; Tung, Y.C.; Sud, S.; Taichman, R.S.; Pienta, K.J.; Takayama, S. Microfluidic system for formation of PC-3 prostate cancer co-culture spheroids. *Biomaterials* **2009**, *30*, 3020–3027. [PubMed]
34. Patra, B.; Chen, Y.H.; Peng, C.C.; Lin, S.C.; Lee, C.H.; Tung, Y.C. A microfluidic device for uniform-sized cell spheroids formation, culture, harvesting and flow cytometry analysis. *Biomicrofluidics* **2013**, *7*, 54114. [CrossRef]
35. Mehta, G.; Hsiao, A.Y.; Ingram, M.; Luker, G.D.; Takayama, S. Opportunities and challenges for use of tumor spheroids as models to test drug delivery and efficacy. *J. Control Release* **2012**, *164*, 192–204. [PubMed]
36. Kato, Y.; Iwamoto, M.; Koike, T.; Suzuki, F.; Takano, Y. Terminal differentiation and calcification in rabbit chondrocyte cultures grown in centrifuge tubes: Regulation by transforming growth factor beta and serum factors. *Proc. Natl. Acad. Sci. USA* **1988**, *85*, 9552–9556. [PubMed]
37. Anada, T.; Fukuda, J.; Sai, Y.; Suzuki, O. An oxygen-permeable spheroid culture system for the prevention of central hypoxia and necrosis of spheroids. *Biomaterials* **2012**, *33*, 8430–8441. [PubMed]
38. Jahn, K.; Richards, R.G.; Archer, C.W.; Stoddart, M.J. Pellet culture model for human primary osteoblasts. *Eur. Cell Mater.* **2010**, *20*, 149–161. [PubMed]
39. Giovannini, S.; Diaz-Romero, J.; Aigner, T.; Heini, P.; Mainil-Varlet, P.; Nesic, D. Micromass co-culture of human articular chondrocytes and human bone marrow mesenchymal stem cells to investigate stable neocartilage tissue formation in vitro. *Eur. Cell Mater.* **2010**, *20*, 245–259. [CrossRef]
40. Ruedel, A.; Hofmeister, S.; Bosserhoff, A.K. Development of a model system to analyze chondrogenic differentiation of mesenchymal stem cells. *Int. J. Clin. Exp. Pathol.* **2013**, *6*, 3042–3048. [PubMed]
41. Ivascu, A.; Kubbies, M. Rapid generation of single-tumor spheroids for high-throughput cell function and toxicity analysis. *J. Biomol. Screen* **2006**, *11*, 922–932. [PubMed]
42. Landry, J.; Bernier, D.; Ouellet, C.; Goyette, R.; Marceau, N. Spheroidal aggregate culture of rat liver cells: Histotypic reorganization, biomatrix deposition, and maintenance of functional activities. *J. Cell Biol.* **1985**, *101*, 914–923. [CrossRef] [PubMed]
43. Friedrich, J.; Ebner, R.; Kunz-Schughart, L.A. Experimental anti-tumor therapy in 3-D: Spheroids—Old hat or new challenge? *Int. J. Radiat. Biol.* **2007**, *83*, 849–871. [CrossRef] [PubMed]
44. Friedrich, J.; Seidel, C.; Ebner, R.; Kunz-Schughart, L.A. Spheroid-based drug screen: Considerations and practical approach. *Nat. Protoc.* **2009**, *4*, 309–324.
45. Froehlich, K.; Haeger, J.D.; Heger, J.; Pastuschek, J.; Photini, S.M.; Yan, Y.; Lupp, A.; Pfarrer, C.; Mrowka, R.; Schleussner, E.; et al. Generation of multicellular breast cancer tumor spheroids: Comparison of different protocols. *J. Mammary Gland. Biol. Neoplasia* **2016**, *21*, 89–98. [PubMed]
46. Keller, G.M. In vitro differentiation of embryonic stem cells. *Curr. Opin. Cell Biol.* **1995**, *7*, 862–869. [PubMed]
47. Tung, Y.C.; Hsiao, A.Y.; Allen, S.G.; Torisawa, Y.S.; Ho, M.; Takayama, S. High-throughput 3D spheroid culture and drug testing using a 384 hanging drop array. *Analyst* **2011**, *136*, 473–478. [CrossRef]
48. Sandu, I.; Fleaca, C.T. The influence of gravity on the distribution of the deposit formed onto a substrate by sessile, hanging, and sandwiched hanging drop evaporation. *J. Colloid. Interface Sci.* **2011**, *358*, 621–625. [PubMed]
49. Lin, B.; Miao, Y.; Wang, J.; Fan, Z.; Du, L.; Su, Y.; Liu, B.; Hu, Z.; Xing, M. Surface tension guided hanging-drop: Producing controllable 3D spheroid of high-passaged human dermal papilla cells and forming inductive microtissues for hair-follicle regeneration. *ACS Appl. Mater. Interfaces* **2016**, *8*, 5906–5916. [CrossRef]
50. Frey, O.; Misun, P.M.; Fluri, D.A.; Hengstler, J.G.; Hierlemann, A. Reconfigurable microfluidic hanging drop network for multi-tissue interaction and analysis. *Nat. Commun.* **2014**, *5*, 4250. [PubMed]
51. Hsiao, A.Y.; Tung, Y.C.; Kuo, C.H.; Mosadegh, B.; Bedenis, R.; Pienta, K.J.; Takayama, S. Micro-ring structures stabilize microdroplets to enable long term spheroid culture in 384 hanging drop array plates. *Biomed. Microdevices* **2012**, *14*, 313–323. [CrossRef] [PubMed]
52. Cavnar, S.P.; Salomonsson, E.; Luker, K.E.; Luker, G.D.; Takayama, S. Transfer, imaging, and analysis plate for facile handling of 384 hanging drop 3D tissue spheroids. *J. Lab Autom.* **2014**, *19*, 208–214. [CrossRef]
53. Nyberg, S.L.; Hardin, J.; Amiot, B.; Argikar, U.A.; Remmel, R.P.; Rinaldo, P. Rapid, large-scale formation of porcine hepatocyte spheroids in a novel spheroid reservoir bioartificial liver. *Liver Transpl.* **2005**, *11*, 901–910. [CrossRef] [PubMed]

54. Lazar, A.; Mann, H.J.; Remmel, R.P.; Shatford, R.A.; Cerra, F.B.; Hu, W.S. Extended liver-specific functions of porcine hepatocyte spheroids entrapped in collagen gel. *Cell. Dev. Biol. Anim.* **1995**, *31*, 340–346. [CrossRef] [PubMed]

55. Ingram, M.; Techy, G.B.; Saroufeem, R.; Yazan, O.; Narayan, K.S.; Goodwin, T.J.; Spaulding, G.F. Three-dimensional growth patterns of various human tumor cell lines in simulated microgravity of a nasa bioreactor. *Cell. Dev. Biol. Anim.* **1997**, *33*, 459–466.

56. Khaoustov, V.I.; Darlington, G.J.; Soriano, H.E.; Krishnan, B.; Risin, D.; Pellis, N.R.; Yoffe, B. Induction of three-dimensional assembly of human liver cells by simulated microgravity. *Cell. Dev. Biol. Anim.* **1999**, *35*, 501–509. [CrossRef] [PubMed]

57. Sebastian, A.; Buckle, A.M.; Markx, G.H. Tissue engineering with electric fields: Immobilization of mammalian cells in multilayer aggregates using dielectrophoresis. *Biotechnol. Bioeng.* **2007**, *98*, 694–700.

58. Okochi, M.; Matsumura, T.; Honda, H. Magnetic force-based cell patterning for evaluation of the effect of stromal fibroblasts on invasive capacity in 3D cultures. *Biosens. Bioelectron.* **2013**, *42*, 300–307.

59. Wang, T.; Green, R.; Nair, R.R.; Howell, M.; Mohapatra, S.; Guldiken, R.; Mohapatra, S.S. Surface acoustic waves (SAW)-based biosensing for quantification of cell growth in 2D and 3D cultures. *Sensors* **2015**, *15*, 32045–32055. [CrossRef] [PubMed]

60. Jaganathan, H.; Gage, J.; Leonard, F.; Srinivasan, S.; Souza, G.R.; Dave, B.; Godin, B. Three-dimensional in vitro co-culture model of breast tumor using magnetic levitation. *Sci. Rep.* **2014**, *4*, 6468. [PubMed]

61. Haisler, W.L.; Timm, D.M.; Gage, J.A.; Tseng, H.; Killian, T.C.; Souza, G.R. Three-dimensional cell culturing by magnetic levitation. *Nat. Protoc.* **2013**, *8*, 1940–1949. [CrossRef] [PubMed]

62. Tseng, H.; Gage, J.A.; Raphael, R.M.; Moore, R.H.; Killian, T.C.; Grande-Allen, K.J.; Souza, G.R. Assembly of a three-dimensional multitype bronchiole coculture model using magnetic levitation. *Tissue. Eng. Part C Methods* **2013**, *19*, 665–675. [PubMed]

63. Souza, G.R.; Molina, J.R.; Raphael, R.M.; Ozawa, M.G.; Stark, D.J.; Levin, C.S.; Bronk, L.F.; Ananta, J.S.; Mandelin, J.; Georgescu, M.M.; et al. Three-dimensional tissue culture based on magnetic cell levitation. *Nat. Nanotechnol.* **2010**, *5*, 291–296.

64. Bazou, D.; Kuznetsova, L.A.; Coakley, W.T. Physical enviroment of 2-D animal cell aggregates formed in a short pathlength ultrasound standing wave trap. *Ultrasound Med. Biol.* **2005**, *31*, 423–430. [CrossRef] [PubMed]

65. Liu, J.; Kuznetsova, L.A.; Edwards, G.O.; Xu, J.; Ma, M.; Purcell, W.M.; Jackson, S.K.; Coakley, W.T. Functional three-dimensional HepG2 aggregate cultures generated from an ultrasound trap: Comparison with HepG2 spheroids. *J. Cell Biochem.* **2007**, *102*, 1180–1189. [PubMed]

66. Chen, P.; Guven, S.; Usta, O.B.; Yarmush, M.L.; Demirci, U. Biotunable acoustic node assembly of organoids. *Adv. Healthc. Mater.* **2015**, *4*, 1937–1943. [PubMed]

67. Guo, F.; Mao, Z.; Chen, Y.; Xie, Z.; Lata, J.P.; Li, P.; Ren, L.; Liu, J.; Yang, J.; Dao, M.; et al. Three-dimensional manipulation of single cells using surface acoustic waves. *Proc. Natl. Acad. Sci. USA* **2016**, *113*, 1522–1527. [CrossRef] [PubMed]

68. Chen, K.; Wu, M.; Guo, F.; Li, P.; Chan, C.Y.; Mao, Z.; Li, S.; Ren, L.; Zhang, R.; Huang, T.J. Rapid formation of size-controllable multicellular spheroids via 3D acoustic tweezers. *Lab Chip* **2016**, *16*, 2636–2643. [CrossRef]

69. Schultz, K.M.; Furst, E.M. High-throughput rheology in a microfluidic device. *Lab Chip* **2011**, *11*, 3802–3809. [PubMed]

70. Yu, L.; Chen, M.C.; Cheung, K.C. Droplet-based microfluidic system for multicellular tumor spheroid formation and anticancer drug testing. *Lab Chip* **2010**, *10*, 2424–2432. [PubMed]

71. Chen, W.; Kim, J.H.; Zhang, D.; Lee, K.H.; Cangelosi, G.A.; Soelberg, S.D.; Furlong, C.E.; Chung, J.H.; Shen, A.Q. Microfluidic one-step synthesis of alginate microspheres immobilized with antibodies. *J. R. Soc. Interface* **2013**, *10*, 20130566. [PubMed]

72. Tehranirokh, M.; Kouzani, A.Z.; Francis, P.S.; Kanwar, J.R. Microfluidic devices for cell cultivation and proliferation. *Biomicrofluidics* **2013**, *7*, 51502. [CrossRef] [PubMed]

73. Hu, G.; Li, D. Three-dimensional modeling of transport of nutrients for multicellular tumor spheroid culture in a microchannel. *Biomed. Microdevices* **2007**, *9*, 315–323.

74. Wu, L.Y.; Di Carlo, D.; Lee, L.P. Microfluidic self-assembly of tumor spheroids for anticancer drug discovery. *Biomed. Microdevices* **2008**, *10*, 197–202.

75. Wang, W.H.; Zhang, Z.L.; Xie, Y.N.; Wang, L.; Yi, S.; Liu, K.; Liu, J.; Pang, D.W.; Zhao, X.Z. Flow-focusing generation of monodisperse water droplets wrapped by ionic liquid on microfluidic chips: From plug to sphere. *Langmuir* **2007**, *23*, 11924–11931. [CrossRef] [PubMed]

76. Christopher, G.; Bergstein, J.; End, N.; Poon, M.; Nguyen, C.; Anna, S.L. Coalescence and splitting of confined droplets at microfluidic junctions. *Lab Chip* **2009**, *9*, 1102–1109. [CrossRef] [PubMed]

77. Kim, L.; Toh, Y.C.; Voldman, J.; Yu, H. A practical guide to microfluidic perfusion culture of adherent mammalian cells. *Lab Chip* **2007**, *7*, 681–694. [PubMed]

78. Agastin, S.; Giang, U.B.; Geng, Y.; Delouise, L.A.; King, M.R. Continuously perfused microbubble array for 3D tumor spheroid model. *Biomicrofluidics* **2011**, *5*, 24110. [PubMed]

79. Toh, Y.C.; Zhang, C.; Zhang, J.; Khong, Y.M.; Chang, S.; Samper, V.D.; van Noort, D.; Hutmacher, D.W.; Yu, H. A novel 3D mammalian cell perfusion-culture system in microfluidic channels. *Lab Chip* **2007**, *7*, 302–309. [CrossRef]

80. Okuyama, T.; Yamazoe, H.; Mochizuki, N.; Khademhosseini, A.; Suzuki, H.; Fukuda, J. Preparation of arrays of cell spheroids and spheroid-monolayer cocultures within a microfluidic device. *J.Biosci. Bioeng.* **2010**, *110*, 572–576. [CrossRef] [PubMed]

81. Chen, S.Y.C.; Hung, P.J.; Lee, P.J. Microfluidic array for three-dimensional perfusion culture of human mammary epithelial cells. *Biomed. Microdevices* **2011**, *13*, 753–758. [PubMed]

82. Sakai, Y.; Hattori, K.; Yanagawa, F.; Sugiura, S.; Kanamori, T.; Nakazawa, K. Detachably assembled microfluidic device for perfusion culture and post-culture analysis of a spheroid array. *Biotechnol. J.* **2014**, *9*, 971–979. [CrossRef]

83. McMillan, K.S.; Boyd, M.; Zagnoni, M. Transitioning from multi-phase to single-phase microfluidics for long-term culture and treatment of multicellular spheroids. *Lab Chip* **2016**, *16*, 3548–3557. [PubMed]

84. Ruppen, J.; Cortes-Dericks, L.; Marconi, E.; Karoubi, G.; Schmid, R.A.; Peng, R.; Marti, T.M.; Guenat, O.T. A microfluidic platform for chemoresistive testing of multicellular pleural cancer spheroids. *Lab Chip* **2014**, *14*, 1198–1205. [CrossRef] [PubMed]

85. Wong, K.H.; Chan, J.M.; Kamm, R.D.; Tien, J. Microfluidic models of vascular functions. *Annu. Rev. Biomed. Eng.* **2012**, *14*, 205–230. [CrossRef] [PubMed]

86. Zhu, Y.; Fang, Q. Analytical detection techniques for droplet microfluidics—A review. *Anal. Chim. Acta* **2013**, *787*, 24–35. [PubMed]

87. Basova, E.Y.; Foret, F. Droplet microfluidics in (bio)chemical analysis. *Analyst* **2015**, *140*, 22–38.

88. Oliveira, A.F.; Pessoa, A.C.; Bastos, R.G.; de la Torre, L.G. Microfluidic tools toward industrial biotechnology. *Biotechnol. Prog.* **2016**, *32*, 1372–1389. [PubMed]

89. Wen, N.; Zhao, Z.; Fan, B.; Chen, D.; Men, D.; Wang, J.; Chen, J. Development of droplet microfluidics enabling high-throughput single-cell analysis. *Molecules* **2016**, *21*, 881. [CrossRef] [PubMed]

90. Martino, C.; deMello, A.J. Droplet-based microfluidics for artificial cell generation: A brief review. *Interface Focus* **2016**, *6*, 20160011. [CrossRef] [PubMed]

91. Thorsen, T.; Maerkl, S.J.; Quake, S.R. Microfluidic large-scale integration. *Science* **2002**, *298*, 580–584. [PubMed]

92. Clausell-Tormos, J.; Lieber, D.; Baret, J.C.; El-Harrak, A.; Miller, O.J.; Frenz, L.; Blouwolff, J.; Humphry, K.J.; Köster, S.; Duan, H. Droplet-based microfluidic platforms for the encapsulation and screening of mammalian cells and multicellular organisms. *Chem. Biol.* **2008**, *15*, 427–437. [CrossRef]

93. Utada, A.; Lorenceau, E.; Link, D.; Kaplan, P.; Stone, H.; Weitz, D. Monodisperse double emulsions generated from a microcapillary device. *Science* **2005**, *308*, 537–541.

94. Utada, A.; Chu, L.-Y.; Fernandez-Nieves, A.; Link, D.; Holtze, C.; Weitz, D. Dripping, jetting, drops, and wetting: The magic of microfluidics. *MRS Bull.* **2007**, *32*, 702–708.

95. Cristini, V.; Renardy, Y. Scalings for droplet sizes in shear-driven breakup: Non-microfluidic ways to monodisperse emulsions. *Fluid Dyn. Mater. Process* **2006**, *2*, 77–94.

96. Renardy, Y. The effects of confinement and inertia on the production of droplets. *Rheol. Acta* **2007**, *46*, 521–529.

97. Baret, J.C. Surfactants in droplet-based microfluidics. *Lab Chip* **2012**, *12*, 422–433.

98. Tumarkin, E.; Kumacheva, E. Microfluidic generation of microgels from synthetic and natural polymers. *Chem. Soc. Rev.* **2009**, *38*, 2161–2168. [PubMed]

99. Um, E.; Lee, D.S.; Pyo, H.B.; Park, J.K. Continuous generation of hydrogel beads and encapsulation of biological materials using a microfluidic droplet-merging channel. *Microfluid. Nanofluidics* **2008**, *5*, 541–549. [CrossRef]

100. Tendulkar, S.; Mirmalek-Sani, S.H.; Childers, C.; Saul, J.; Opara, E.C.; Ramasubramanian, M.K. A three-dimensional microfluidic approach to scaling up microencapsulation of cells. *Biomed. Microdevices* **2012**, *14*, 461–469. [PubMed]

101. Chan, H.F.; Zhang, Y.; Leong, K.W. Efficient one-step production of microencapsulated hepatocyte spheroids with enhanced functions. *Small* **2016**, *12*, 2720–2730.

102. Orive, G.; Hernandez, R.M.; Gascon, A.R.; Calafiore, R.; Chang, T.M.; de Vos, P.; Hortelano, G.; Hunkeler, D.; Lacik, I.; Shapiro, A.M.; et al. Cell encapsulation: Promise and progress. *Nat. Med.* **2003**, *9*, 104–107.

103. Eun, Y.J.; Utada, A.S.; Copeland, M.F.; Takeuchi, S.; Weibel, D.B. Encapsulating bacteria in agarose microparticles using microfluidics for high-throughput cell analysis and isolation. *ACS Chem. Biol.* **2010**, *6*, 260–266. [CrossRef]

104. Zamora-Mora, V.; Velasco, D.; Hernández, R.; Mijangos, C.; Kumacheva, E. Chitosan/agarose hydrogels: Cooperative properties and microfluidic preparation. *Carbohydr. Polym.* **2014**, *111*, 348–355. [CrossRef]

105. Mahadik, B.P.; Haba, S.P.; Skertich, L.J.; Harley, B.A. The use of covalently immobilized stem cell factor to selectively affect hematopoietic stem cell activity within a gelatin hydrogel. *Biomaterials* **2015**, *67*, 297–307.

106. Velasco, D.; Tumarkin, E.; Kumacheva, E. Microfluidic encapsulation of cells in polymer microgels. *Small* **2012**, *8*, 1633–1642. [CrossRef] [PubMed]

107. Nicodemus, G.D.; Bryant, S.J. Cell encapsulation in biodegradable hydrogels for tissue engineering applications. *Tissue Eng. Part B Rev.* **2008**, *14*, 149–165.

108. Hsu, M.N.; Luo, R.; Kwek, K.Z.; Por, Y.C.; Zhang, Y.; Chen, C.H. Sustained release of hydrophobic drugs by the microfluidic assembly of multistage microgel/poly (lactic-co-glycolic acid) nanoparticle composites. *Biomicrofluidics* **2015**, *9*, 052601. [CrossRef]

109. Karnik, R.; Gu, F.; Basto, P.; Cannizzaro, C.; Dean, L.; Kyei-Manu, W.; Langer, R.; Farokhzad, O.C. Microfluidic platform for controlled synthesis of polymeric nanoparticles. *Nano. Lett.* **2008**, *8*, 2906–2912. [CrossRef] [PubMed]

110. Shum, H.C.; Kim, J.W.; Weitz, D.A. Microfluidic fabrication of monodisperse biocompatible and biodegradable polymersomes with controlled permeability. *J. Am. Chem. Soc.* **2008**, *130*, 9543–9549.

111. Choi, C.H.; Jung, J.H.; Rhee, Y.W.; Kim, D.P.; Shim, S.E.; Lee, C.S. Generation of monodisperse alginate microbeads and in situ encapsulation of cell in microfluidic device. *Biomed. Microdevices* **2007**, *9*, 855–862.

112. Akbari, S.; Pirbodaghi, T. Microfluidic encapsulation of cells in alginate particles via an improved internal gelation approach. *Microfluid. Nanofluidics* **2014**, *16*, 773–777.

113. Kim, M.S.; Yeon, J.H.; Park, J.K. A microfluidic platform for 3-dimensional cell culture and cell-based assays. *Biomed. Microdevices* **2007**, *9*, 25–34. [CrossRef]

114. Tamura, M.; Yanagawa, F.; Sugiura, S.; Takagi, T.; Sumaru, K.; Kanamori, T. Click-crosslinkable and photodegradable gelatin hydrogels for cytocompatible optical cell manipulation in natural environment. *Sci. Rep.* **2015**, *5*, 15060.

115. Wang, Y.; Wang, J. Mixed hydrogel bead-based tumor spheroid formation and anticancer drug testing. *Analyst* **2014**, *139*, 2449–2458.

116. Shi, X.; Sun, L.; Jiang, J.; Zhang, X.; Ding, W.; Gan, Z. Biodegradable polymeric microcarriers with controllable porous structure for tissue engineering. *Macromol. Biosci.* **2009**, *9*, 1211–1218.

117. Wang, J.; Cheng, Y.; Yu, Y.; Fu, F.; Chen, Z.; Zhao, Y.; Gu, Z. Microfluidic generation of porous microcarriers for three-dimensional cell culture. *ACS Appl. Mater. Interfaces* **2015**, *7*, 27035–27039. [CrossRef]

118. Li, T.; Zhao, L.; Liu, W.; Xu, J.; Wang, J. Simple and reusable off-the-shelf microfluidic devices for the versatile generation of droplets. *Lab Chip* **2016**, *16*, 4718–4724.

119. Suzuki, K.; Homma, H.; Murayama, T.; Fukuda, S.; Takanobu, H.; Miura, H. Electrowetting-based actuation of liquid droplets for micro transportation systems. *J.Adv. Mech. Des. Syst. Manuf.* **2010**, *4*, 365–372. [CrossRef]

120. Pollack, M.G.; Fair, R.B.; Shenderov, A.D. Electrowetting-based actuation of liquid droplets for microfluidic applications. *Appl. Phys. Lett.* **2000**, *77*, 1725–1726. [CrossRef]

121. Barbulovicnad, I.; Yang, H.; Park, P.S.; Wheeler, A.R. Digital microfluidics for cell-based assays. *Lab Chip* **2008**, *8*, 519–526.

122. Cho, S.K.; Moon, H.; Kim, C.J. Creating, transporting, cutting, and merging liquid droplets by electrowetting-based actuation for digital microfluidic circuits. *J. Microelectromech. Syst.* **2003**, *12*, 70–80.

123. Griffth, E.J.; Akella, S.; Goldberg, M.K. Performance characterization of a reconfigurable planar array digital microfluidic system. In *Design Automation Methods and Tools for Microfluidics-Based Biochips*; Springer: Berlin, Germany, 2006; pp. 329–356.

124. Yoon, J.Y.; Garrell, R.L. Preventing biomolecular adsorption in electrowetting-based biofluidic chips. *Anal. Chem.* **2003**, *75*, 5097–5102.

125. Hong, J.; Kim, Y.K.; Won, D.J.; Kim, J.; Lee, S.J. Three-dimensional digital microfluidic manipulation of droplets in oil medium. *Sci. Rep.* **2015**, *5*, 10685.

126. Fiddes, L.K.; Luk, V.N.; Au, S.H.; Ng, A.H.; Luk, V.; Kumacheva, E.; Wheeler, A.R. Hydrogel discs for digital microfluidics. *Biomicrofluidics* **2012**, *6*, 14112–1411211. [CrossRef]

127. Au, S.H.; Chamberlain, M.D.; Mahesh, S.; Sefton, M.V.; Wheeler, A.R. Hepatic organoids for microfluidic drug screening. *Lab Chip* **2014**, *14*, 3290–3299. [CrossRef]

128. Aijian, A.P.; Garrell, R.L. Digital microfluidics for automated hanging drop cell spheroid culture. *J. Lab Autom.* **2015**, *20*, 283–295. [CrossRef]

129. Vadivelu, R.K.; Ooi, C.H.; Yao, R.Q.; Velasquez, J.T.; Pastrana, E.; Diaz-Nido, J.; Lim, F.; Ekberg, J.A.; Nguyen, N.T.; St John, J.A. Generation of three-dimensional multiple spheroid model of olfactory ensheathing cells using floating liquid marbles. *Sci. Rep.* **2015**, *5*, 15083.

130. Sarvi, F.; Arbatan, T.; Chan, P.P.Y.; Shen, W. A novel technique for the formation of embryoid bodies inside liquid marbles. *Rsc. Adv.* **2013**, *3*, 14501–14508.

131. Oliveira, N.M.; Correia, C.R.; Reis, R.L.; Mano, J.F. Liquid marbles for high-throughput biological screening of anchorage-dependent cells. *Adv. Healthc. Mater.* **2015**, *4*, 264–270.

132. Ooi, C.H.; Van Nguyen, A.; Evans, G.M.; Gendelman, O.; Bormashenko, E.; Nguyen, N.T. A floating self-propelling liquid marble containing aqueous ethanol solutions. *RSC Adv.* **2015**, *5*, 101006–101012.

133. Khaw, M.K.; Ooi, C.H.; Mohd-Yasin, F.; Vadivelu, R.; John, J.S.; Nguyen, N.T. Digital microfluidics with a magnetically actuated floating liquid marble. *Lab Chip* **2016**, *16*, 2211–2218. [CrossRef]

134. Han, X.; Lee, H.K.; Lim, W.C.; Lee, Y.H.; Phan-Quang, G.C.; Phang, I.Y.; Ling, X.Y. Spinning liquid marble and its dual applications as microcentrifuge and miniature localized viscometer. *ACS Appl. Mater. Interfaces* **2016**, *8*, 23941–23946. [CrossRef]

135. Castro, J.O.; Neves, B.M.; Rezk, A.R.; Eshtiaghi, N.; Yeo, L.Y. Continuous production of Janus and composite liquid marbles with tunable coverage. *ACS Appl. Mater. Interfaces* **2016**, *8*, 17751–17756.

136. Enger, J.; Goksör, M.; Ramser, K.; Hagberg, P.; Hanstorp, D. Optical tweezers applied to a microfluidic system. *Lab Chip* **2004**, *4*, 196–200. [CrossRef]

137. Grier, D.G. A revolution in optical manipulation. *Nature* **2003**, *424*, 810–816. [CrossRef]

138. Chiou, P.Y.; Chang, Z.; Wu, M.C. Droplet manipulation with light on optoelectrowetting device. *J. Microelectromech. Syst.* **2008**, *17*, 133–138.

139. Park, S.Y.; Chiou, P.Y. Light-driven droplet manipulation technologies for lab-on-a-chip applications. *Adv. OptoElectron.* **2011**, *2011*.

140. Baigl, D. Photo-actuation of liquids for light-driven microfluidics: State of the art and perspectives. *Lab Chip* **2012**, *12*, 3637–3653.

141. Jiang, D.; Park, S.Y. Light-driven 3D droplet manipulation on flexible optoelectrowetting devices fabricated by a simple spin-coating method. *Lab Chip* **2016**, *16*, 1831–1839.

142. Pandiyan, V.P.; John, R. Optofluidic bioimaging platform for quantitative phase imaging of lab on a chip devices using digital holographic microscopy. *Appl. Opt.* **2016**, *55*, A54–59.

143. Shin, D.; Daneshpanah, M.; Anand, A.; Javidi, B. Optofluidic system for three-dimensional sensing and identification of micro-organisms with digital holographic microscopy. *Opt. Lett.* **2010**, *35*, 4066–4068.

144. Jagannadh, V.K.; Adhikari, J.V.; Gorthi, S.S. Automated cell viability assessment using a microfluidics based portable imaging flow analyzer. *Biomicrofluidics* **2015**, *9*, 024123. [CrossRef]

145. Moldovan, N.I.; Hibino, N.; Nakayama, K. Principles of the *Kenzan* method for robotic cell spheroid-based 3D bioprinting. *Tissue Eng. Part B Rev.* **2016**. [CrossRef]

146. Mehesz, A.N.; Brown, J.; Hajdu, Z.; Beaver, W.; da Silva, J.V.; Visconti, R.P.; Markwald, R.R.; Mironov, V. Scalable robotic biofabrication of tissue spheroids. *Biofabrication* **2011**, *3*, 025002.

147. Kudan, Y.V.; Pereira, F.D.; Parfenov, V.A.; Kasyanov, V.A.; Khesuani, Y.D.; Bulanova, Y.A.; Mironov, V.A. spreading of tissue spheroids from primary human fibroblasts on the surface of microfibrous electrospun polyurethane matrix (a scanning electron microscopic study). *Morfologiia* **2015**, *148*, 70–74.

148. Whatley, B.R.; Li, X.; Zhang, N.; Wen, X. Magnetic-directed patterning of cell spheroids. *J. Biomed. Mater. Res. A* **2014**, *102*, 1537–1547. [CrossRef]

149. Leonard, F.; Godin, B. 3D in vitro model for breast cancer research using magnetic levitation and bioprinting method. *Methods Mol. Biol.* **2016**, *1406*, 239–251.

150. Susienka, M.J.; Wilks, B.T.; Morgan, J.R. Quantifying the kinetics and morphological changes of the fusion of spheroid building blocks. *Biofabrication* **2016**, *8*, 045003. [CrossRef]

151. Munaz, A.; Vadivelu, R.K.; John, J.A.; Nguyen, N.T. A lab-on-a-chip device for investigating the fusion process of olfactory ensheathing cell spheroids. *Lab Chip* **2016**, *16*, 2946–2954.

152. Maschmeyer, I.; Lorenz, A.K.; Schimek, K.; Hasenberg, T.; Ramme, A.P.; Hubner, J.; Lindner, M.; Drewell, C.; Bauer, S.; Thomas, A.; et al. A four-organ-chip for interconnected long-term co-culture of human intestine, liver, skin and kidney equivalents. *Lab.Chip* **2015**, *15*, 2688–2699.

153. Esch, M.B.; Smith, A.S.; Prot, J.M.; Oleaga, C.; Hickman, J.J.; Shuler, M.L. How multi-organ microdevices can help foster drug development. *Adv. Drug Deliv. Rev.* **2014**, *69–70*, 158–169.

154. Materne, E.M.; Ramme, A.P.; Terrasso, A.P.; Serra, M.; Alves, P.M.; Brito, C.; Sakharov, D.A.; Tonevitsky, A.G.; Lauster, R.; Marx, U. A multi-organ chip co-culture of neurospheres and liver equivalents for long-term substance testing. *J. Biotechnol.* **2015**, *205*, 36–46.

155. Bhatia, S.N.; Ingber, D.E. Microfluidic organs-on-chips. *Nat. Biotechnol.* **2014**, *32*, 760–772. [CrossRef]

156. Materne, E.M.; Maschmeyer, I.; Lorenz, A.K.; Horland, R.; Schimek, K.M.; Busek, M.; Sonntag, F.; Lauster, R.; Marx, U. The multi-organ chip—A microfluidic platform for long-term multi-tissue coculture. *J. Vis. Exp.* **2015**, e52526.

157. Kretzschmar, K.; Clevers, H. Organoids: Modeling development and the stem cell niche in a dish. *Dev. Cell* **2016**, *38*, 590–600.

158. Schepers, A.; Li, C.; Chhabra, A.; Seney, B.T.; Bhatia, S. Engineering a perfusable 3D human liver platform from iPS cells. *Lab Chip* **2016**, *16*, 2644–2653.

159. Saenz, D.B.L.; Compte, M.; Aceves, M.; Hernandez, R.M.; Sanz, L.; Alvarez-Vallina, L.; Pedraz, J.L. Microencapsulation of therapeutic bispecific antibodies producing cells: Immunotherapeutic organoids for cancer management. *J. Drug Target* **2015**, *23*, 170–179.

160. Yang, Y.; Opara, E.C.; Liu, Y.; Atala, A.; Zhao, W. Microencapsulation of porcine thyroid cell organoids within a polymer microcapsule construct. *Exp. Biol. Med.* **2017**, *242*, 286–296.

161. Khademhosseini, A.; Eng, G.; Yeh, J.; Kucharczyk, P.A.; Langer, R.; Vunjak-Novakovic, G.; Radisic, M. Microfluidic patterning for fabrication of contractile cardiac organoids. *Biomed. Microdevices* **2007**, *9*, 149–157.

162. Shen, Y.; Hou, Y.; Yao, S.; Huang, P.; Yobas, L. In vitro epithelial organoid generation induced by substrate nanotopography. *Sci. Rep.* **2015**, *5*, 9293.

MDPI AG

St. Alban-Anlage 66

4052 Basel, Switzerland

Tel. +41 61 683 77 34

Fax +41 61 302 89 18

http://www.mdpi.com

Micromachines Editorial Office

E-mail: micromachines@mdpi.com

http://www.mdpi.com/journal/micromachines

www.ingramcontent.com/pod-product-compliance
Lightning Source LLC
Chambersburg PA
CBHW051858210326
41597CB00033B/5942